NATURAL HISTORY AND THE INDIAN ARMY

"Bounded by themselves, and unregardful
In what state God's other works may be,
In their own tasks all their powers pourings;
These attain the mighty life you see"

– Mathew Arnold

NATURAL HISTORY AND THE INDIAN ARMY

J.C. Daniel

Lieut General Baljit Singh, AVSM, VSM (Retd.)

BOMBAY NATURAL HISTORY SOCETY

OXFORD UNIVERSITY PRESS
MUMBAI DELHI KOLKATA CHENNAI

Oxford University Press, Walton Street, Oxford OX2 6DP

Oxford, New York,
Athens, Auckland, Bangkok,
Calcutta, Cape Town, Chennai, Dar-es-Salaam,
Delhi, Florence, Hong Kong, Istanbul,
Karachi, Kuala Lumpur, Madrid, Melbourne,
Mexico City, Mumbai, Nairobi, Paris,
Singapore,Taipei, Tokyo, Toronto,
and associated companies in
Berlin, Ibadan

© Bombay Natural History Society, 2009

COMPILED AND EDITED BY: J.C. DANIEL AND LIEUT GEN BALJIT SINGH

Layout and Design: V. Gopi Naidu
Cover Photo of Leopard: Vivek Sinha
Back cover Photo : Orange-bellied Leafbird *Chloropsis hardwickii*

ISBN (10): 0-19-806450-0
ISBN (13): 9780198064503

The book is available at:
Bombay Natural History Society
Hornbill House,
Opp. Lion Gate, Shaheed Bhagat Singh Road,
Mumbai 400 001, Maharashtra, India.
Telephone: 0091-022-2282 1811, Fax: 0091-022-2283 7615
Email: bnhs@bom4.vsnl.net.in
Websites: www.bnhs.org

Bombay Natural History Society in India is registered under Bombay Public Trust Act 1950: F244 (Bom) dated 6th July 1953

PROCESSED BY TRENDZ PHOTOTYPESETTERS, Email: gotrendz@gmail.com
PRINTED BY SPECIFIC ASSIGNMENTS INDIA PVT. LTD., Email: parag@specificassignment.com

PUBLISHED BY THE BOMBAY NATURAL HISTORY SOCIETY, HORNBILL HOUSE, OPP. LION GATE, SHAHEED BHAGAT SINGH ROAD, MUMBAI 400 001 AND CO-PUBLISHED BY MANZAR KHAN, OXFORD UNIVERSITY PRESS, YMCA LIBRARY BUILDING, 1 JAI SINGH ROAD, NEW DELHI 110 001.

ACKNOWLEDGEMENTS

This book is a tribute to the Soldier Naturalists of the late 17th, 18th and early 20th centuries. We are grateful to Lieut Gen R.K. Gaur and his publishers Brijbasi Printers Pvt. Ltd. for permission to reproduce excerpts from his book INDIAN BIRDS.

The assistance rendered by Ms Vibhuti Dedhia, General Manager, Publications, BNHS, and her team is gratefully acknowledged. The Director, BNHS, and the Publications Subcommittee of BNHS are thanked for their assistance in producing this publication. We are grateful to Mr. Sachin Kulkarni for administrative assistance and the Librarian, BNHS, Mrs Nirmala Reddy and Library Assistants Tarendra Singh and Sadanand Shirsat for their help.

Special thanks are due to Mr. V. Gopi Naidu for the excellent design.

We are particularly grateful to Mr. Vivek Sinha, Meethil Momaya, A.J.T. Johnsingh, Isaac Kehimkar, Ashok Kumar, Ashok Captain, Theodore Baskaran, Patrick David, Rajat Bhargava, Shubhalaxmi Vaylure, P. Jegannathan, Sudheer Agashe, Nikhil Bhopale, Ajay Desai, Usha Lachungpa, Kumaran Sathasivam, Kedar Bhide, Varad Giri, Ravi Singh and Mechanised Infantry Regimental Centre for their superb photographs, we are grateful to Ms. Rivka Israel for editorial assistance.

In the United Service Institute of India, New Delhi, we are grateful to Maj Gen Y.K. Gera, Maj Gen P.J.S. Sandhu, Sqn Ldr R.T.S. Chhina and Librarian Mr. P.K. Varma, who helped us promptly with data to formulate brief biographical notes on contributors of the article in this compilation.

CONTENTS

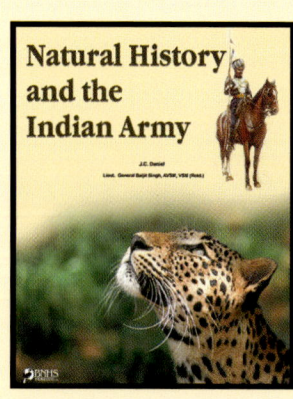

Foreword

It is a remarkable fact that in the early years, the study of the natural history of the Indian subcontinent was more or less the exclusive preserve of officers of the Indian Army. Beginning with the fading years of the 18th century, some of the natural history inclined officers of the Indian Army studied flora and fauna, particularly fauna, and became experts thereon. Their findings were recorded in various journals in the U.K., but mainly in the *Journal of the Bombay Natural History Society* after 1886 which year saw the first issue of the *Journal*. This compendium of articles by officers of the Indian Army is more or less exclusively of their publications in the Society's *Journal*. It was also the Army's singular honour to have drawn the attention of the Indian Government to the dire need for the conservation of the wildlife of the country. The pamphlet which was prepared and sent to the Indian Government through the Society by Lt Col Burton, and which in due course saw the formation of the Indian Board for Wildlife, is the 25th and final article in this collection. This book is a record and a tribute to the remarkable personality of the Indian Army.

It was with great pleasure and much satisfaction that I went through the copy of this book compiled by J.C. Daniel and Lt Gen Baljit Singh on the natural history contributions of Indian Army Officers over more than two hundred years of the Army's existence. It is a truly wonderful story, a story we should be proud of and wish to carry forward as a matter both of curiosity and duty. Names like Hardwicke and Tickell, Jerdon and Wall, Swinhoe and Evans, Kirtikar and Jayakar are well known to every natural history scientist in the world . We owe them a debt of gratitude. It was time that their pioneering labours were recorded for posterity. The authors have done just that through this compilation of articles by some of these soldier amateur-naturalists of yore. It is a very readable record which, I believe, should be widely read by Officers of our armed forces both Regular and Paramilitary. In its honourable service the Army fought on the field for the security of this country. It also fought a battle to conserve the country's wonderful, almost incomparable, natural heritage spread over a zoogeographical area over thirty degrees of latitude encompassing almost every type of climate met with in the world. The officers found their recreation in field sport. Concurrently many of them found time in their leisure hours to pursue the serious study of Wildlife of India. They became amateur scientists, many of them were acknowledged and even applauded for the quality of their research and investigation in their lifetimes.

I strongly believe that the three Services, and the paramilitary forces too, have an important role to play in carrying forward this work as their contribution to conserve national biological wealth, for the well-being of generations yet to be born.

M.P. AWATI
P.V.S.M., Vr.C.
Vice Admiral (Retd.)

NATURAL HISTORY
AND
THE INDIAN ARMY

A tribute to the remarkable personality of the Indian Army

EXPLORATION AND DOCUMENTATION OF INDIAN NATURAL HISTORY: CONTRIBUTIONS BY INDIAN ARMY OFFICERS 1778-2002

A RETROSPECTIVE OF 225 YEARS

By Lieut Gen Baljit Singh, A.V.S.M., V.S.M. (Indian Army) (Retd.)

The Indian Army's heritage of valour on the battlefield is a byword in the annals of world military histories. But the Indian Army's equally distinguished and scholarly heritage in the field of India's Natural History, remains generally unknown and unsung.

John Keay, a present day writer of travel and history, sums up best the pioneering role of Indian army officers in the exploration and documentation of India's fauna and flora thus:

"The men who discovered India came as amateurs; by profession they were soldiers and administrators. But they returned home as giants of scholarship."

YEARS OF INVESTIGATION AND DOCUMENTATION 1778-1947

The scope of this narrative is limited exclusively to the contributions made by Indian army officers to the study of Indian natural history. And at the outset it is conceded that the officers of the Indian Civil Service (ICS), the Imperial Police in India (IP), the Indian Forest Service (IFS) and other allied Services contributed concurrently as much and more to the study and documentation of India's natural history.

The Phenomenon of Indian Army Officer-Naturalist

From its very beginnings in the eighteenth century, the Army in India was an exclusive service. The uniform, the ceremonial and its conduct on the field, together wove an edifying image. The groomed appearance, the dash, the chivalry, courteous manners and impeccable character placed the army officer in an unique class. Little wonder

Lieut Gen Baljit Singh (Retd.) served with distinction in the Indian Army for over 36 years (1956-1992) and was awarded the A.V.S.M. and V.S.M. Concomittantly he strove to promote conservation of wildlife and nature as a way of life within and by the Armed Forces. He was for a time a Trustee of the World Wide Fund for Nature - India.

therefore, that a high place in the passing-out merit list from Sandhurst was the first criterion for the young man who hoped to be commissioned into the Indian Army in preference to the British Army.

A significant feature of the Indian Army of those days was that an officer was expected to combine business with pleasure, without detracting from his devotion to duty. And if the path of duty led the army officer where he could also indulge his creative urges, he would be considered foolish not to spend his spare time agreeably. This was the liberal milieu which aided and encouraged many army officers to gain recognition as men of contemporary merit in pursuits way outside the ambit of soldiering. A few among such individuals even carved a permanent niche in the world of Art and Science and left behind enduring works which became definitive or classics for generations to follow. So it was that among those army officers who were attracted to the study of natural history of India, a good many would merit entry to a Hall of Fame.

Maj. Gen. Thomas Hardwicke who had arrived in India in 1778 as a cadet in the Bengal Artillery, was the first to establish in India the discipline of scientific investigation of natural history. He was not just the leader of the pack of Indian army officer-naturalists who followed in his wake but he was the pioneer and trend-setter in this field for India as a whole. When in December 1823 Gen. Hardwicke retired and set sail for London, after 45 years of distinguished army service in India, he had on the sidelines of his army career made a monumental contribution by any yardstick to the study of Natural History in India. If his name seldom appears in Indian natural history retrospectives it is chiefly because he explored the natural world in India a century before the Bombay Natural History Society (BNHS) was even born.

The army officer-naturalists came from two streams of literacy. Gen. Hardwicke and his kind who joined the Army at age 20 to 22 presumably had school education only. They were self-taught, on-the-job natural history investigators. A few of this class, such as Gen. Hardwicke, Col. Tickell, Col. Bingham and Brig. Evans most probably were gifted with intuitive genius in this field. Then there were others, Maj. Jerdon, Maj. Hingston, Col. Kirtikar, Col. Sir Chopra and Col. Wall who were post-graduates in medical sciences possibly with exposure to botany and zoology. I believe they were men with cultivated scientific temper who joined the ranks of army officers through the Indian Medical Service. Just a few, such as Col. Swinhoe, MA (Oxon) were in a class of their own. But all of them were driven by one common purpose outside their chosen profession, and that was the single-point focus on the exploration and documentation of Indian natural history.

Maj. Gen. Thomas Hardwicke

Discovering Birds of India: 1778-1897

It would appear that the study of Indian natural history began with its bird species. Gen. Thomas Hardwicke lit the flame. From his arrival at Fort William in Calcutta in 1778, till he retired in 1823, he was mostly stationed at Barrackpore, Dum Dum, Fatehgarh and Kanpur. For two years (1789-90) he also operated around Mysore and once he took a month's leave which he spent in the Garhwal region. His focus lay in collecting bird specimens, their nests and eggs and in preserving bird skins. This he did personally and among his prizes was the first specimen of the White-crested Laughingthrush. In addition, he employed a shikari on his household staff who was tasked and trained to add new specimens to his collection. To enlarge the area of his collection he enthused friends to add to his trove. So, Lieut Counsel stationed at Almora sent him the first specimens of the Koklass and the Cheer Pheasants and three species of Jays. Similarly the Hon. Edward Gardner and Dr. Wallich sent him from Nepal the first specimens of a number of birds

White-crested Laughingthrush
Garrulax leucolophus

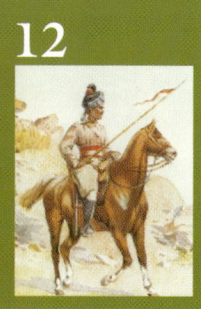

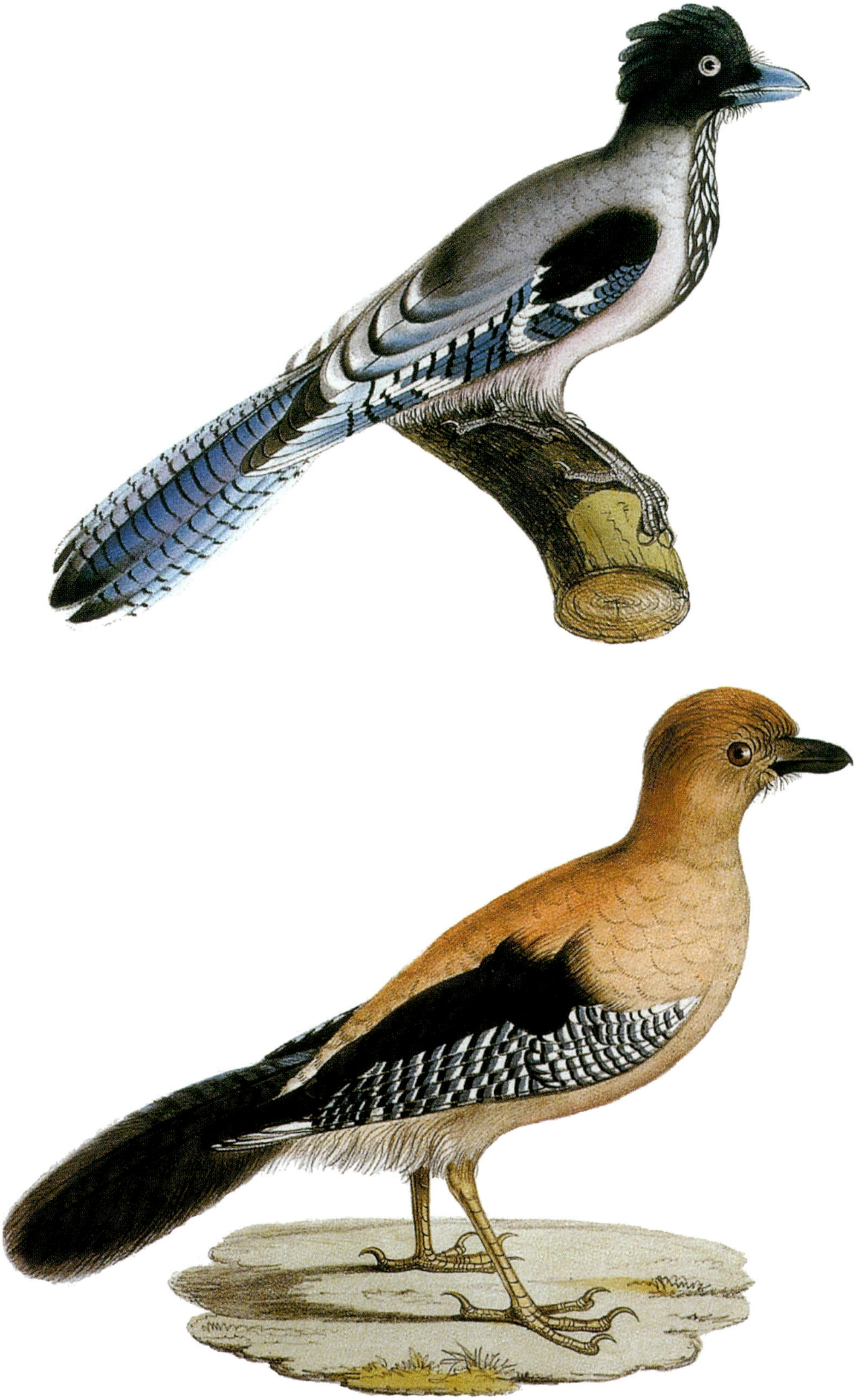

(HARDWICKE COLLECTION)

Black-headed Jay *Garrulus lanceolatus* and Eurasian Jay *Garrulus glandarius*
From Hardwicke and Gray ILLUSTRATIONS OF INDIAN ZOOLOGY

including the Blood Pheasant. Sir Norman Kinnear, the Director of the British Natural History Museum, London and previously the first Curator of the BNHS acknowledges that "by far the largest (collection) was the one made by Maj. Gen. Hardwicke."

In the absence of taxidermy and photography, Hardwicke employed the most talented artists of Bengal to paint, draw and sketch the bird species from his collection. By 1802, this bird art collection, together with paintings of mammals and insects, made up 32 large and bulging folio volumes. By about the same time, he seems to have given the finishing touches to his field notes on birds.

Hardwicke wrote the first descriptions of the White-crested Laughingthrush in 1815, of the Blood Pheasant in 1821 and of the Cheer Pheasant in 1827. In 1811, on leave in London, he made over his bird paintings and notes to Dr. Latham who was writing the first scientific descriptions of certain species of Indian birds. Sad to say, the *Chloropsis hardwickii*" is the only memorial to his labours.

Hardwicke bequeathed his entire collection of specimens (birds, mammals, insects, reptiles, butterflies etc.) and paintings to the British Museum which was perhaps the world's first comprehensive acquisition and display of Indian Natural History. This priceless collection was used by J.E. Gray, the curator, to publish ILLUSTRATIONS OF INDIAN ZOOLOGY: CHIEFLY SELECTED FROM THE COLLECTION OF MAJ. GEN. HARDWICKE in two volumes in 1830-35. This is probably the first ever exclusively book published on Indian Natural History. A hundred and sixty seven years later, in 2002, many of the Hardwicke paintings adorned Jagmohan Mahajan's book SPLENDID PLUMAGE.

In the ten-year period following Hardwicke's departure from India, his legacy was taken up with considerable success by Capt. James Franklin of the 1st Bengal Cavalry and Capt. W.H. Sykes of the Bombay Presidency Army. Though Capt. Franklin was an acknowledged geologist, during a journey to study rocks from Calcutta to Saugar via Benares in 1826, he collected 156 species of birds. Franklin gifted the skins, paintings and his notes to the Asiatic Society, Calcutta from where they were sent to the Zoological Society, London.

In October 1824, Capt. Sykes was appointed Statistical Reporter to the Bombay Government. Over the next seven years, on the sidelines of his duty, Sykes collected 236 species of birds and his paper "A Catalogue of the Raptorial and Incessorial Orders observed in the Dukhun (Deccan)" was published by the Zoological Society, London in 1832. Of the birds of the Indian subcontinent today, eleven first descriptions are credited to Capt. Sykes and two birds carry his name.

Orange-bellied Chloropsis
Chloropsis hardwickii

Lieut Col S.R.Tickell's arrival at Calcutta in 1833 to join the 31st Bengal Native Infantry infused further momentum to the study of bird life. Due to exigencies of service he was assigned to the Civil Administration and fortunately for ornithology, to the south-west districts of Bengal which had not been explored by Gen. Hardwicke. His output of the first five years in India was published by the Asiatic Society titled A LIST OF BIRDS COLLECTED IN THE JUNGLES OF BARABHUM AND DHALBUM; in todays' Jharkhand and Orissa. In 1840, Tickell followed with a fresh list in the maiden number of the *Calcutta Magazine of Natural History*.

Tickell enlarged his area of bird study up to Darjeeling. He had prepared the manuscript of a book complete with illustrations by himself but death claimed him too soon in 1866. The Minute Book of the Zoological Society London has the entry on 1st Dec. 1874, "it was announced by the Secretary that Col R.S. Tickell, late of H.M.'s Indian Army, had presented to the Society's library a very fine illustrated M.S. work in seven small folio Vols. on the ornithology of India." Sir Norman Kinnear is also on record that "Tickell was one of the best field naturalists India has known." He is credited with eleven first descriptions of Indian birds and four birds fly about with his name.

Franklin's or Savanna Nightjar
Caprimulgus affinis

Lieut Col Samuel Richard Tickell

Maj. T.C. Jerdon, the first of the tribe of medico army officer-naturalists, who arrived in India in 1835, hard on the heels of Tickell, was to become one of the greatest boons to Indian ornithology. Tickell and Jerdon's concurrent but independent efforts of the next 30 years were to give Indian ornithology a sound foundation. Fortunately, Jerdon was assigned to the Madras Presidency so that for the next 16 years while he operated in what is today Andhra Pradesh, Karnataka, Kerala and Tamil Nadu, his contemporary Tickell was active in Orissa, Bengal, Bangladesh, Assam, Jharkhand, Bihar and Darjeeling. Within the first four years, Jerdon made a handsome collection of bird skins and published his first work CATALOGUE OF THE BIRDS OF THE INDIAN PENINSULA serialised by the *Madras Journal of Literature and Science* between 1839-41. Two supplements followed in 1845-46 bringing the total birds to 420 species in his catalogue. Jerdon next completed his work ILLUSTRATIONS OF INDIAN ORNITHOLOGY with descriptive text in 1846.

Between 1846 and 1858 Jerdon was perhaps more preoccupied with the reptiles and mammals of India. But this decade of Jerdon's absence was admirably filled by Capts. Pemberton and W.E. Boyes. Pemberton was sent on a mission by the Government to Bhutan in 1836, accompanied by a botanist and taxidermist. He collected 500 bird skins representing 126 species inhabiting Bhutan. No paper was written on this collection, Government enterprises being the same then as now (!), but most of them were recorded by Ludlow in "Birds of Bhutan" in the *Ibis*, published a century later in 1937.

Capt. Boyes of the 6th Cavalry had made a considerable collection of birds from the Himalayas, Uttar Pradesh and Rajasthan; the latter not having been covered by any army officer thus far. Details of Boyes, collection are not known, but when he died in 1854, it was auctioned and mostly purchased by one Dr. Wilson who gifted it to the Philadelphia Museum*. Some of Capt. Boyes, notes were used in the descriptive text to John Gould's BIRDS OF ASIA Vol. VII published in 1883.

Tickell's Flycatcher *Cyornis tickelliae*

Jerdon in 1854 was posted to the 4th Light Cavalry at Saugar, and in 1857 he operated with his regiment in Central India. Evidently this enlarged his sphere of birds to Madhya Pradesh and Uttar Pradesh. His major break came after 1857. Convalescing from illness at Darjeeling, he had the opportunity to meet the Viceroy of India, Lord Canning. Jerdon seized the chance and presented to the Viceroy his dream of compiling books on birds and mammals of India. Lord Canning was obviously impressed by both Jerdon's sincerity of purpose and the professional merit of the project. He was moved post-haste to Fort William and for the next six years was allowed to travel extensively to the Punjab, Kashmir and all the hill stations of India, gathering specimens and field knowledge for his project. The result was Jerdon's BIRDS OF INDIA in three Volumes in 1862-64 covering the vast area from the Himalayas to Kanyakumari and from the Indus to the Brahmaputra.

It may not be too rash to claim that this book was the first definitive benchmark in the establishment and pursuit of scientific ornithology in India. That working single-handed Jerdon covered 1,008 species of India's birds out of around 1,300 now recorded, remains

---

*On a request from me, Mr. Aasheesh Pittie most enthusiastically obtained a confirmation on January 22, 2007 from Nathan H. Rice, Ornithology Collection Manager, at the Academy of Natural Sciences, Philadelphia, that they indeed have 328 specimens of Capt Boyes' collection held by them.

Maj. Thomas Caverhill Jerdon

an unparalleled achievement in the annals of ornithology. He is credited with 14 first descriptions of Indian birds and five of them tag his name to their wings. With the re-discovery of Jerdon's Courser in 1986 by BNHS, Jerdon's memory was once again revived in India.

Delivering the Sálim Ali Memorial Lecture in 1996, J.C. Daniel said of Jerdon: "As a field naturalist he was not to be equalled till Sálim Ali came on the scene." Finally, there is a footnote to "The Roll of Indian Medical Service" which sums up Jerdon thus: "he lived and served before the days of decorations." Not just that but he even used up all his personal resources in the pursuit of natural history so that when he died, to the shock of his wife and two children, they were rendered insolvent.

Col Swinhoe joined the Bombay Staff Corps in 1858, when Col Tickell and Maj Jerdon were at their productive best. Luckily, Swinhoe was to serve with the Army of the Indus and was in Afghanistan for the first 25 years of his service in areas comparatively less explored. He collected 70 bird skins from Sind and gifted them to the BNHS which perhaps gave a start to the Society's collection. During the Second Afghan War, he also collected three specimens of the Coronetted Sandgrouse from the battlefield of Maiwund near Kandahar in February 1881. These specimens are at the British Museum and still carry Swinhoe's collector's label in original. He is credited with eleven first descriptions of Indian birds and three are named after him.

The last significant contribution to field ornithology by an army officer in the 19th century was made by Lieut Col J.W. Yerbury. Commissioned in the Royal Artillery, he was later transferred to the Political Department and stationed at Aden. He was to become an authority on Aden's natural history. His papers on the 'Birds of Aden' were published in *Ibis* in 1886 and in *JBNHS* in 1897.

Jerdon's Courser
Rhinoptilus bitorquatus

Mammals of India: 1800-1936

Indian army officers contributed handsomely to the study of mammalian fauna at two levels, directly and indirectly. In the first category are the likes of Gen Hardwicke, Capt Sykes, Col Tickell, Maj Jerdon and Col Jayakar who compiled species-specific texts and laid the foundations for the documentation of mammals of India. And the second and the much larger category was essentially made up of sportsmen-naturalists whose combined observations of their quarry in the field provided the scientists valuable insights into habitats and animal behaviour to build up a composite profile of each species. The most prolific in this category were Brig Gen Burton and Cols Fenton, Burton, Mosse, Stockley and Ward and a host more whose names keep tumbling out when one has the patience and recourse to archival material.

Gen. Hardwicke was also the pioneer of mammalogy in India. He described the Goral and the Indian Gerbil between 1800-1810, which may well be the first for any Indian mammal. In the normal course, Hardwicke gifted some mammals from his collection to the British Museum, the Zoological and the Linnean Societies. The latter Society belied Hardwicke's trust as they forestalled his first descriptions of the Gaur and the Four-horned Antelope. This Society continued to ignore his contributions because as per their Minute Book for 1821, a description of the Panda was communicated by Hardwicke, read in his absence but never published.

Coronetted or Crowned Sandgrouse
Pterocles coronatus

Indian Gerbil *Tatera indica*

KEDAR BHIDE

Spiny-tailed Lizard *Uromastyx hardwickii*

Again in 1823, his papers on the Tailless Deer and *Ovis argali* (Mountain Sheep) met the same fate. Was this professional envy to keep Hardwicke away from the scientists, turf? All his descriptions were accompanied by paintings from his priceless collection. J.C. Daniel's latest book on Reptiles and Amphibians has two colourful and handsome lizards named *Eublepharis hardwickii* (Leopard Gecko) and *Uromastyx hardwickii* (Spiny-tailed Lizard), a miserly tribute to an Indian army officer-naturalist who is perhaps an all time "great".

Capt W.H. Sykes arrived at Bombay in 1824, a year after Hardwicke's retirement. Over the next seven years while working on the statistical report of the Deccan Plateau he also made documentaries on mammals. In 1831, he published in the Proceedings of the Zoological Society a paper describing the Indian Gazelle, the Wild Dog, and listing 36 other species with brief accounts of their habits and distribution.

VIVEK SINHA

Hangul *Cervus elaphus hanglu*

Hawk-Eagle *Nisaetus* sp. from Jerdon's ILLUSTRATIONS OF INDIAN ORNITHOLOGY (1846)

Malabar White-headed Starling *Sturnia blythii* from Jerdon's ILLUSTRATIONS OF INDIAN ORNITHOLOGY (1846)

In 1838, Lieut Thomas Hutton of the 37th Native Infantry described and named the Sind Markhor and the Urial. He also added new insights to the knowledge on the Sind Ibex. After the First Afghan War, he published "Rough Notes on the Zoology of Afghanistan" in the *Journal of the Asiatic Society of Bengal*, in 1854.

The first book on Indian mammals was written by Col Tickell supported by his own illustrations, possibly in 1850, but for reasons unknown it was never published. The manuscript and illustrations lie preserved in the library of the Zoological Society, London. Nevertheless, Tickell published regularly in the *Calcutta Magazine of Natural History* which includes papers on the Sloth Bear, Brown Flying Squirrel and Anteater. His account of the habits of Gibbons was considered a classic. When in 1888 W.T. Blandford published his book on mammals he used Tickell's manuscript extensively.

Capt (?) A.L. Adams was the medical officer with the 22nd Foot, the Cheshire Regiment, and between 1849-56 served at Poona, Karachi and Rawalpindi and made expeditions to the Himalayas and Kashmir. His paper on the mammals he had observed in these areas was published by the Zoological Society in 1858.

Hoolock Gibbon *Hylobates* hoolock

To Maj Jerdon goes the credit again of the first book published on Indian mammals. He travelled and researched extensively for three years till in 1867 emerged the book, MAMMALS OF INDIA covering 247 species. This was to remain the standard work for the next 21 years till it was replaced by Blandford's volumes. Blandford acknowledges having used Jerdon's notes. And to quote J.C. Daniel again: "Jerdon's field notes were impeccable and have stood the test of time and are in many instances the only information that we have on some of the rare species."

Lieut Col A.S.G. Jayakar was appointed assistant surgeon in the Sultanate of Oman at Muscat in 1867. He was a keen collector and in the process discovered three goat-like skulls which on investigation resulted in the discovery of a new species, the Arabian Tahr, which was later named after its discoverer, as *Hemitragus jayakarii*.

Of the sportsmen-naturalists, Brig Gen R.G. Burton's was a prominent name. His two books on the Tiger (1933 and 1936) and a paper titled "Some Natural History Notes on Tiger" in the *JBNHS* were significant. His three notes on the Panther (*JBNHS* Vols. 17 to 26) were also interesting. His younger brother Col R.W. Burton who followed him to India in 1889, was perhaps the single largest contributor to the *JBNHS* in its long history. His paper "A History of Shikar in India" (Vol. 50) is a masterly summation of the natural history of the big Cats, Crocodiles, Sloth Bear, Hyena, sheep and goats. And his article "The Indian Wild Dog" (Vol. 41) remains relevant even today.

Arabian Tahr *Hemitragus jayakarii*

During the first ten years of the 20th century (1895-1910) significant knowledge of the mammals of Hazara and the North-West Frontier Province was gathered by Col Magrath, Maj Dunn and Capt Whitehead. They also enriched the collection of both the British Museum and of the Society. Capt Walton IMS a surgeon with Col Younghusband's expedition to Lhasa, collected mammals on the Tibetan plateau but unfortunately no report was published.

One of the happiest fields for the sportsman-naturalist was Kashmir. Col. Stockley was the first to compile detailed notes on sporting mammals. But the task of collecting Kashmir mammals was taken up earnestly by Col. A.E. Ward between 1924-1929. Ward was to send specimens to the British Museum from time to time, and he published in the *JBNHS* identifications of the specimens sent to London. Regrettably, no consolidated account of his work was ever published.

Then there was Lieut Col A.H. Mosse. Though a graduate from Sandhurst, he spent almost his entire service with great distinction in the Political Department. His empathy with the Panther and knowledge of its habits and distribution were most significant.

Lieut Col A.S.G. Jayakar

This class of sportsmen-naturalists was best summed up by the late Sálim Ali: "their unquenchable thirst for scientific enquiry, contributed so significantly to what may be termed …the knowledge of the Natural History of the animals they hunted."

Snakes of India: 1850-1930

Again, it was Maj Jerdon who first amassed an impressive collection of snakes. In 1853 he published the "Catalogue of Reptiles Inhabiting Southern India" in the *Journal of the Asiatic Society of Bengal*. Most of the species described are today in the British Museum, but his drawings are lost. By 1870 Jerdon had made new discoveries and enlarged the collection which was published as NOTES ON INDIAN HERPETOLOGY by the Asiatic Society. But to the loss of Indian natural history, Jerdon's book on reptiles did not see light of day. He had handed the completed manuscript to the press in 1868, and proofs reached England, but no one knows what happened next. Having retired from service in February 1868, this great officer-naturalist fell ill at Guwahati, reached London in 1870, but did-not recover and passed away on 12th June 1872. J.C. Daniel holds that "Jerdon was an extraordinarily versatile scientist who was equally at home in the study of mammals, birds or reptiles".

Almost 50 years were to elapse before Col. Frank Wall, IMS, emerged as an authority on Indian snakes. His texts are precise and elaborate and the coloured drawings made by himself are so excellent that J.C. Daniel writing in THE BOOK OF INDIAN REPTILES AND AMPHIBIANS in 2002 acknowledges that "all colour drawings of the snakes in the Book are credited to Col Wall's originals". Wall shared his collection equally between the British Museum and the BNHS. But his scientific text "A Popular Treatise on the Common Indian Snakes" in 20 parts and illustrated by coloured plates and diagrams was published exclusively in the *JBNHS*.

Lieut Col K.G. Gharpurey I.M.S. a contemporary of Wall, was also an accomplished herpetologist. Having seen active service in Somalia, Africa and Persia, he was to spend the last 15 years of his service in the Bombay Presidency where he indulged fully in the study of snakes. Between 1927-34 he published six papers on snakes in the *JBNHS*. In 1935 he authored the book "The Snakes of India" which ran into at least six editions and remained the standard text for the next 25 years.

ISAAC KEHIMKAR

Russell's Viper *Daboia russelii*

Bird Photography

Love of birds and bird photography are usually interdependent and often one interest leads on to the other. So it was with Col. R.S.P. Bates who arrived in India in 1920. His first love was birds and between 1923 and 1939 he published 17 papers in the *JBNHS* on bird behaviour covering areas from Chittagong in the east to Kashmir in the north-west on such diverse subjects as "Do Birds Employ Ants to Rid themselves of Ectoparasites", "On the Parasitic Habits of the Pied Crested Cuckoo," and "Communal Nest Feeding in Babblers". But Bates is remembered most for pioneering bird photography in India. His popular series "Birds Nesting with a Camera" in the *JBNHS* was to later form the basis of his book BIRD LIFE IN INDIA in 1930. Probably this was the first bird photography book of India. Bates travelled extensively in India, watching and photographing birds. He left India in 1947 but in 1952 co-authored with his friend from the Indian Railways E.H.N. Lowther the book BREEDING BIRDS OF KASHMIR. Bates, works were exhibited at the International Exhibition of Nature Photography in London in 1950 where he was awarded a bronze plaque. Sálim Ali was to write about Bates: "An ardent lover of birds and a knowledgable and painstaking field ornithologist, he made significant contributions to Indian ornithology… Many of his portraits of Indian birds must still rank amongst the finest ever made." Coming from Sálim Ali that is a very high rating of merit.

Botany of India: 1880 to 1942

All contributors in the field of botany were medicos from the IMS. The trend-setter was non other than the indefatigable Maj. Jerdon. Regrettably, just as he was sprouting out as a botanist, first the exigencies of Army service and next ill health checked him in mid-stride. So, Lieut Col K.R. Kirtikar led the field. He saw active service in the Second Afghan War with distinction and spent the residual years of his service in the Bombay Presidency. He made a large and significant collection of dried plants, watercolours of algae and fungi, and notes on botancial objects. He wrote extensively for the *JBNHS*. He published POISONOUS PLANTS OF BOMBAY in 1893 but his magnum opus was INDIAN MEDICINAL PLANTS in two volumes published posthumously in 1920.

Col Sir R.N. Chopra, IMS was a patriot who wanted to use India's botanical wealth to make the nation self-reliant in pharmaceutial drugs. On the basis of his research and field knowledge, he stated that "Nearly three fourths of the drugs used in the pharmacopoeias of the World, grow in a state of Nature in Jammu and Kashmir and as many as 42 essential oil-bearing plants are grown in J&K." In 1933, he published his paper in the INDIAN MEDICAL JOURNAL on *Rauwolfia serpentina* which was then to become the wonder drug plant of the world. He also authored two books INDIGENOUS DRUGS OF INDIA: MEDICINAL AND POISONOUS PLANTS in two Volumes and GLOSSARY OF INDIAN MEDICINAL PLANTS which remain the standard works to date. He was the only Indian in Army ranks to be knighted, for work related to Natural History.

Maj. Gen. Sir Arthur Cotton KCSI* commissioned in the Corps of Engineers arrived at Madras in 1821. For the next ten years as he soldiered in Burma and north-east India he also made the first comprehensive collection of orchids of the north-east. Unfortunately the whereabouts of this collection are not known though one orchid is named after Gen Cotton. As an engineer, he is remembered for designing and constructing the first anicuts over the rivers Godavari, Krishna and Cauvery between 1836-64, which are functional to date.

Cerbera thevatia. Nat. Ord. Apocynaceae from Kirtikar's POISONOUS PLANTS OF BOMBAY

Fishes of India: 1848 to 1946

The first scientific paper on fish was written in 1848, "On the Freshwater Fish of South India" and authored by none else but Maj Jerdon. It was Jerdon's ambition to complete his quartet on natural history with a volume on fish. But death intervened and Sir Walter Elliot wrote "To no one is Indian Science so deeply indebted as to Dr. Jerdon, not for his discoveries considerable as they were, but for enabling others to follow his steps." And so it was that Lieut Col Arthur MacDonald, who studied the sporting fish, wrote extensively on the subject in the *JBNHS*. This work was ultimately published in book form as CIRCUMVENTING THE MAHSEER AND OTHER SPORTING FISH IN INDIA AND BURMA. Though limited in scope, the work became a classic for its times.

There were two other army officers who studied and wrote comprehensively on the sporting fishes of India. Capt C.W.W. Conway published his findings in the book SUNLIT WATERS in 1938. It must have been a great success as it appeared in its second edition the very next month. A masterly treatise on fishes and fishing, written in the manner of the 16th century classic by Izzak Walton, was THE COMPLETE INDIAN ANGLER by Col John Masters published in 1942. It is not just natural history but a great literary work in its class. Both are embellished with exquisite watercolours, pen-and-ink sketches and photographs of fish and fishing locales, making them collectors, items today.

Goonch *Bagarius bagarius*

*Knight Commander of the Star of India

Great Orange Tip

Orange Oakleaf (dorsal)

Orange Oakleaf (ventral)

Plain Tiger

Great Orange Tip

Danaid Eggfly

Butterflies of India: 1880 to 1956

In a period of 40 years between 1836 and 1876 were born five men, Swinhoe, Yerbury, Bingham, Tytler and Evans, all of whom came to be commissioned to the Indian Army. Between them they explored and mapped out the entire Lepidoptera of India. It is not surprising, therefore, that after ornithology, the discipline of natural history which was most comprehensively studied and scientifically documented was Lepidoptera. Col. Charles Swinhoe set the lead and attained international status in his life-time. The Indian Government entrusted to him the task of completing that magnificent work on butterflies LEPIDOPTERA INDICA. The chapters on the "Blues", the "Skippers" and the "Whites" were compiled entirely by Swinhoe. However, it was the "Moths" section in which he was to excel both for the breadth of his knowledge and the sheer size of his collection. For, he had collected 40,000 specimens comprising 7,000 different species. Of these, 400 species were described by Swinhoe for the first time. In 1922, the last year of his life, he completed the book A REVISION OF THE GENERA OF THE FAMILY LIPARIDAE covering some 1,130 detailed entries. He was universally recognised as an expert on all matters connected with Lepidoptera and was the recepient of many international honours for his services to entomology. Col Swinhoe was also one of the eight founding members of the BNHS.

Lieut Col Yerbury had spent the last 10-15 years of his service at Aden. He wrote frequentely in the *JBNHS* on Aden's butterflies and on the art of collecting gad-flies, bot-flies and warble-flies and became an authority on this group of insects.

A contemporary of Swinhoe and an equally gifted Lepidopterist was Lieut Col C.T. Bingham. Though an officer of the Bengal Staff Corps, he spent the better part of his service in Burma and retired as the Chief Conservator of Forests, there. By the time BNHS came into being, Bingham shifted his focus from birds to insects and published 14 papers in Vols. 3 to 14 of the *JBNHS*. His collection became the basis for much of the FAUNA OF BRITISH INDIA volumes on butterflies. On the death of Dr. Blandford, Col. Bingham assumed responsibility and produced the three volumes dealing with Nymphalids, Papilionids and Pieris.

Maj. Gen. Sir H.C. Tytler was commissioned in the Indian Army in 1887. He served in the Naga Hills, Lushai Hills, Manipur, Assam, Sikkim and then right across to the North-West Frontier. Being a keen naturalist, he indulged in collecting and studying butterflies in all these areas. To him is, therefore, attributed the claim that "Sir Harry amassed one of the finest collections of butterflies ever made in India". He was a regular contributor to the *JBNHS,* "adding to the knowledge of Indian butterflies… many new species… and many others till then hardly known". He gifted the rarities from his collection to the British Museum.

Brig W.H. Evans, the last of the big five, was India-born, the son of Gen. Sir Horace Evans, commandant of the 8th Gurkha Regiment at Shillong. Commissioned from Sandhurst to the Corps of Engineers, he won the DSO for gallantry in France in WWI. He was to emerge as the keenest and most productive of all Lepidopterists of not only India but the world. His greatest work, an unrivalled breakthrough in entomology, was the construction of "Keys" for the identification of various species and subspecies of butterflies based on differences in genitalia and later even faeces. Evan's Keys formed the basis for the KEYS OF TALBOT'S BUTTERFLIES OF INDIA, CEYLON AND BURMA. Evans serialised articles in the *JBNHS* which were published in book form in 1927, including a large number of species and subspecies newly named by him. The second edition of this book included three new genera and 161 new species and subspecies of Hesperiidae and 52 new species and subspecies of other families.

After retirement in 1931, Brig Evans worked as an Hon. Associate at the British Museum. He worked exclusively and single-handedly on the HESPERIIDAE OF THE WORLD

published by the Museum in several volumes. The first volume dealt with Africa, the second with Europe, Asia and Australia and the next four volumes with North and South Americas. To gauge the expanse of this work take the volume on Europe, Asia and Australia which alone had 502 pages, dealing with 1,641 active specific and subspecific names for the area. Despite failing health, damaged lungs during WWI, a bad heart and bad knees, Evans got up the steps to his office in the Museum and worked almost till the last day of his life. But no work was important if anyone came seeking help and advice on butterflies.

This account would be incomplete without the mention of one butterfly which carries the name of that inimitable naturalist of India, Gen Hardwicke. J.E. Gray, the then curator of the British Museum who had exclusive access to General Hardwicke's notes and collections, named the Common Blue Apollo ("a creamy white Swallowtail, dusted with black scales, spots of crimson, blue and dusky black...") *Parnassius hardwickei* Gray.

Conservation of Wild Life: 1890 to 1948

In the closing decade of the 19th century the large body of sportsmen-naturalists were disturbed to find that the wildlife of India was dwindling at an alarming rate. They were also quick to reason that natural habitats across the country were coming under severe pressure of the combined forces of ever increasing human and cattle populations and of the emerging trends for rapid growth of commerce and industry. The first to sound the alarm was Lieut Col L.L. Fenton around 1895. Fenton was commissioned into the Royal Artillery in 1870 but spent most of his 21 years, service as Political Assistant at Kathiawar. He was literally thrown to the lions, and the knowledge of the Asiatic lion he so acquired he shared through frequent writings in the *JBNHS*. Fenton was to realise how isolated and land-locked these Asiatic lions of the Gir Forest were and how vulnerable to extinction through epidemic and inbreeding. Fenton laboured hard to arouse people's consciousness of the need to conserve this species and his was perhaps the first voice of the conservation movement per se on the subcontinent.

Common Blue Apollo
Parnassius hardwickei

Very soon almost all sportsmen-naturalists were to espouse Fenton's cause concerning practically all game animals and birds. However, their collective concerns could not be channelised into action as the two World Wars scattered these officers all over the world. Luckily, one among them, Lieut Col R.W. Burton was to spend his entire service in India. Commissioned from Sandhurst in 1880, his promising career was cut short by a riding accident in 1903. He was permanently crippled and was absorbed in the Cantonment Magistrates Service. He was one of India's foremost sportsmen-naturalists but he is remembered more as a great proponent of the conservation movement. The summation of his efforts was the pamphlet "Wild Life Preservation: India's Vanishing Asset" published in *JBNHS* in 1948, followed by a supplement. The basic thrust was that wildlife and its associated habitats across the country are national assets which must be preserved for posterity. And that it is the duty of the State to do so. The pamphlet was sent to all Secretaries of the Government at Delhi and in the States. Burton persevered till in 1952 the Prime Minister, Pandit Jawaharlal Nehru announced the creation of the Indian Board for Wild Life. Though happy, Burton at the same time was disappointed. The basic pillar of his concept

Tiger in Kanha

was that preservation of wildlife must become a people's movement. Just as the people had been aroused to attain Independence, in like manner they must be encouraged to protect their wildlife. Burton was to meet and seek Mahatma Gandhi's help but the Mahatma was felled by an assassin just days before the meeting. Burton's text of the concept of conservation was so comprehensive that all the texts we see today have little new to suggest.

THE PRESENT:
FROM SPORTSMEN-NATURALISTS TO NATURE CONSERVATION PROMOTERS
1947-2002

Genesis of the Ideology: 1947 to 1964

The post Independence army officer had inherited two important legacies in the field of natural history. The first was the meaningful contribution to the study of fauna of India by the handful of army officer-naturalists and the second, the necessity for the paradigm shift from sportsmen-naturalists to nature conservation promoters. By 1947, the animal world of India had been comprehensively understood and documented. So there was little more that an army officer-Naturalist could contribute in this field. The future seemed to point inexorably to the promotion of nature conservation.

The first twenty years of Independence had the Indian Army committed to wars; first in J&K, then the insurgency in Nagaland, then skirmishes on the Indo-Tibet border leading to war with China from Ladakh to Arunachal and a second war with Pakistan along the western border. As they soldiered on battlefields representing India's varied biogeographic zones, the army officers also glimpsed bits of wildlife and its habitats which their predecessors had helped document painstakingly over the previous 150 years. Many of them must have concluded that here was a national asset which must be preserved for posterity. During this long period of hostilities on our borders (1947-65) the world was experiencing a new movement raised by the voice of reason and sanity to conserve nature and natural resources spearheaded in the West scene by Sir Julian Huxley, Sir Peter Scot and Rachel Carson. In India the lead was given by Dr. Sálim Ali. For imparting the scientific idiom to the conservation movement much of the credit would also be shared by M. Krishnan, Humayun Abdulali, E.P. Gee, Zafar Futehally, the researchers of the BNHS, the Wildlife Institute of India, the late Dr. Anil Aggarwal and Dr. M. S. Swaminathan. Viewed against this backdrop it is not surprising that many Army Officers became proponents of Col. Burton's beliefs for a shift from Sportsmen-Naturalists to Nature Conservation Promoters.

The Formative Years: 1965 to 1985

It is always difficult to peg the moment in time for such a significant paradigm change but in the instant case forces of history have been helpful. During the Raj an annual duck shoot at the Bharatpur wetlands by the Commander-in-Chief and his "guns" had become a compulsory event. Gen J.N. Chaudhary was the last Army Chief to indulge in this legacy on 23rd February 1964. He was succeeded by Gen P.P. Kumaramangalam (decorated for gallantry against Rommel's Panzer Corps, taken prisoner but escaped), a keen naturalist and a visionary who was obsessed with preventing the degradation of India's green cover and consequent soil erosion. He conceived the idea of the Land Army peopled by the youth and led by retired army officers to take up the challenge in the rural countryside. The idea of the Territorial Army Ecological Battalions was conceived

around this time. He encouraged the Army's Shikar Clubs to take on the wider role of Nature Clubs. This writer was the first to seize this opportunity. On 1st October 1965, in celebration of the National Wildlife Week the Shikar Club at Deolali was redesignated School of Artillery Nature Club with the special screening of the MGM film *Serengeti Shall Not Die*. I introduced the audience to the story of the filmmaker Paul Grizemeck who raised funds single-handed so that Serengeti's wildlife may survive but who himself died in the process at the young age of 22. General Kumaramangalam was persuaded to take up Presidentship of WWF-India but he resigned after a brief spell for personal reasons.

The band of army officer enthusiasts encouraged by Gen Kumaramangalam grew in numbers over the next 20 years (1966-85). Many among them had become Brigadiers where they, for the first time, could influence policy making. Luckily for them, Lieut Gen K. Sundarji, the Western Army Commander, was inclined to nature conservation and he soon became the patron-saint of the unofficial conservation promoters, league. He authorised the holding of the first Army Nature Awareness Workshop. Twenty officers (Capts to Lieut Cols) were given exposure to nature for 7 days at Bharatpur and at Sariska Tiger Reserve in March 1985, under Brigs Ashok Verma and Baljit Singh. Mr. H.S. Panwar, Director of the Wildlife Institute, Dehra Dun was among those who spent a day interacting with the officers and imparting training. Henceforth such workshops were to become annual fixtures in the training schedule of the Indian Army as a whole.

When in 1986 Gen Sundarji became the Chief of Army staff he created an Environment Cell under the Military Training Directorate at the Army Headquarters. They were tasked to organise and monitor nature awareness workshops, produce and procure training materials and interact with personnel from NGOs such as the BNHS and WWF-India. The Cell produced a Special Army Training Memorandum which addressed nature awareness training, establishment of Nature Clubs in the Army-run schools and the greening of cantonments and manoeuvre ranges. Gen Sundarji also approved the idea that suitable army officers would address cadets at the Indian Military Academy, Dehra Dun and the National Defence Academy, Khadakvasla, once a year. The Defence Services Staff College, Wellington and the College of Combat, Mhow were later included on the lecture itinerary. I evolved the theme "India's Environmental Concerns and a Role for the Army" and undertook these lectures for about ten years. Maj Gen E D' Souza (Retd.) a BNHS Executive Committee member, was also active in this field. In retrospect it may be stated that these efforts were very useful. More than six middle-rank army officers have since, of their own choosing, graduated from the Home Study Course in Ornithology run by the Institute of Bird Studies and Natural History at the Rishi Valley Education Centre. Two recent additional graduates are Air Marshal (Retd.) Carriappa and Lieut Gen Pankaj Joshi, PVSM, Chief of the Integrated Defence Staff, a recently created apex staff post at the Ministry of Defence.

However, the Chief who formally defined the Army's role in the conservation of nature was Gen. B.C. Joshi. In an Army Headquarters Directive personally signed by him, among other policy he stated, "The profession of Arms meets all threats external or internal, against the Nation. An insidious threat to the environment has come through ecological vandalism. We have to gear up to contend with it. For a start, we must convert all cantonments and Military Stations into model ecological (entities) while at the same time, doing our bit to regenerate degraded ecology. This requires study, imagination and above all, the will to fight." To implement it, the Environment Cell at the Army HQ was restructured and similar Cells were created at each Army Command HQ; lower down the chain of command an existing staff branch was specifically nominated for this task. Perhaps the most pragmatic connected reform at the Army HQ was the restructuring of the Land and Works Directorate into "Land, Works and Environment Directorate" so as to bring the land use policy of the Army in consonance with the ecological imperatives. Gen Joshi entered the Army into a tripartite arrangement with WWF-India and CEE, Ahmedabad, for a start to define "The Role of Indian Army Cantonments, Depots, Manoeuvre Areas and Military Farms as Biodiversity Niches and Habitat Refuges". The concept paper including the action plan were conceived and prepared by this writer. It was presented to Gen. Joshi at the Army HQ in October 1993 and was approved for implementation forthwith. Meanwhile Gen Joshi got the Government to sanction one additional Territorial Army Eco Battalion for Pithoragarh region in Garhwal. He wanted the Army's considerable land holdings at Pithoragarh to be converted into the living green-gene-pool of India's medicinal herbs and plants. But death dealt a severe blow as the Chief died in harness. That his successors to date have on assuming office re-pledged the commitment of the Indian Army to support the conservation movement, is an indication that the idea has taken root for good.

A Few Modest Achievements

It is worth knowing that the Armed Forces have never received any directive from the Government to protect the country's natural environment, not even as an implied hint. That they do so of their own convictions with diligence and through dedication is simply because of their pride in being responsible citizens of the country. Mostly they act in individual capacities on minor conservation-related causes at the local levels. For instance, in 1985, a fenced-in area was to be vacated. There was a herd of Blue Bull inside, which would fall to local shikaris guns within days. Acting in my individual capacity I obtained the Governments (MOEF) permission to translocate the Blue Bull herd to another permanent safe location, on Army premises. The Government also directed the Wildlife Institute of India to tranquilize and assist in the trans-location of the animals. There seems to be no parallel to this project in the country. From conception to completion it took six months to accomplish, and at zero project funding. Two years later, I learnt that the Greyhound coursing lobby in Punjab had had the Common Indian Hare declared a pest under the Wildlife (Protection) Act for a two year period. That would have decimated this species for good in Punjab. I sought the intervention of the Prime Minister of India, convinced him, and had the Punjab Government issue a fresh Gazette Notification cancelling the old one.

At the institutional level, Government and the Army have supported all NGOs; prominent being the BNHS with their Black-neck Crane Project in Ladakh and their Narcondam Hornbill Project in the Andamans, WWF-INDIA with their Bio-diversity Hotspots programmes in Arunachal, Sikkim and the Western Ghats and their TRAFFIC programme relating to Shahtoosh and Snow Leopard pelts. For the MOEF the Army prepared a blueprint for effective radio communications network in Project Tiger Reserves. Lately, the Army cleared the Harike Wetland of Water Hyacinth in a Government-sponsored project. Recently the *Reader's Digest* carried an article on the Rhino in Kaziranga. An inset to the text stated that the Army had created artifical mud hills inside the Sanctuary so that during floods the Rhino need not leave the precints of their safe haven. Going back to the 1970s, Vice Admiral M.P. Awati, as the Commandant of National Defence Academy, Khadakvasla, had successfully exposed many generations of cadets of the Army, Navy and Air Force to nature awareness through guest speakers such as Dr. Sálim Ali, through regular nature camps conducted by the BNHS on the campus. He got the vast wilderness on the NDA campus declared and developed into a Wildlife Sanctuary. With Peafowl and Sambar roaming freely on the campus, the future generations of Armed Forces officers were bound to develop special empathy for wildlife. It is not surprising therefore to read that recently the Coast Guard readily rushed to the Vengurla Rocks in the Arabian Sea to help NGOs free the nests and nesting colonies of the Edible-nest Swiftlets and Marine Gulls from the clutches of the poaching mafia.

Brig. Ranjit Talwar took premature retirement from the Army and joined WWF-India's staff in March 1994. He restructured their TRAFFIC-India wing resulting in sustained seizures of tiger and leopard skins and bones. In 1996 he organised and headed their Tiger Conservation Cell (now designated Tiger & Wildlife Division) where he continues to date. Among other activities, he conceived and has successfully operated the scheme to pay compensation to the aggrieved villagers within 48-72 hours of any cattle being killed by a carnivore.

An incident that I was a witness to on 6th September 2002 at HQ Western Army Command, Chandimandir, best sums up how deep the ethos of nature conservation has seeped into the psyche of many senior army officers. The Chief Wildlife Warden Haryana (CWLW) had been operating a Rescue and Rehablitation Centre for wild animals which stray from the Shivaliks to urban areas. This ten-acre patch in the heart of Chandimandir Cantonment was now to come up for utilisation under the Master Plan. The Army Commander wanted to know where and when the new facility for rescued animals would come up. It transpired that budgeting for the new enclosure might take the CWLW more than a year. And what would happen to stray wild animals in the interim? The CWLW had no hesitation to state that such animals would surely be preyed upon and killed. This was not acceptable to the Army Commander. So he told the CWLW Haryana that HQ Western Command would give him a grant of Rs. 2 lakhs forthwith from their own private welfare funds, and the CWLW gave an undertaking that he would have the new enclosure up in a month. Till then the present facility on Army land would remain operative!

Translocated Nilgai

After this, I had no more doubts that promotion of nature conservation had become a way of life with many army officers. And that they are among those citizens of India who feel inspired by Article 51-A(g) of the Constitution of India which enjoins that each citizen shall:

> "Protect and improve the Natural Environment
> including forests, lakes, rivers and wildlife
> and have compassion for wild creatures."

THE FUTURE

It is for the citizens and nature conservation NGOs, especially the BNHS, to understand the potential of our Armed Forces as nature conservation promoters and influence Government policy to create something like the model which exists in the USA. The Pentagon has a regular department for environmental protection headed by a two-star General (equivalent of our Maj Gen). Among other activities, through a Congressional decree the US Army Engineers are authorised to compete through open tenders as and when any of America's wetlands have to be restored. It is on record that among the best restoration work has been by the US Army Engineers. And almost all manoeuvre and

firing ranges in the USA also double as Wilderness Refuges or National Parks (through a Congressional dispensation), some of which are home to some of America's threatened species like the Prairie Bison and the Bald-headed Eagle to just mention two. Those who are, and will in time to come be in the vanguard of the nature conservation movement, must aim to create a similar enduring facility in the Indian Army before it is too late.

CONCLUSION

At the end of the day, let it suffice to remember that each one of this impassioned band of Indian army officer-naturalists gained entry to what John Keay termed as the most exclusive fraternity of "Giants of Scholarship". They created for the Indian Army a scholarly heritage parallel to the heritage of valour. This too is a fact of history for us to cherish with equal pride.

■ ■ ■

The Strychnine Tree

Strychnos nux-vomica (Linn)

By Lieut Col K R Kirtikar, I.M.S., F.L.S. (Indian Army)
Year of publication: 1894

A tree over 40 feet in height, with a straight thick trunk. *Root*: Thick and with a yellowish epidermis; very bitter. *Stem*: Often 12 feet in circumference. *Branches*: Dense irregular, covered with a smooth ash-coloured bark. Young shoots shining and deep green, often tinged with red; bark nodose, bitter, glabrous. *Petiole*: $^1/_5$–$^1/_2$ inch long; deeply grooved. *Stipules*: None, says Roxburgh in his "Coromandel Plants". Between the opposite leaves there is a raised line, which is perhaps a rudimentary interpetiolar stipule. *Leaves*: Glabrous on both sides, shining, opposite, entire, coriaceous, often decussate in an oblique manner, arising from stout nodes; ovate or rotundate, sometimes elliptically oblong; 3-5 nerved; shortly acuminate or almost apiculate. 1½ x 6 inches; usually 3-5 inches long. Base obtuse, somewhat unequal. *Peduncle*: ½-2 inches. *Flowers*: Many, small, greenish-white, appearing with young leaves on short slender pedicels; collected on small terminal pubescent corymbose cymes 1-2 inches in diameter, at the end of the branchlets or on short axillary shoots; pentamerous; bisexual. *Calyx*: 5-parted, persistent, ¼ or $^1/_5$ the size of the corolla. *Corolla*: Valvate, hypogynous; regular, tubular or funnel-shaped, with a 5-lobed reflexed short limb. Tube ¼ to $^1/_3$ inch long, glabrous at the throat, lobes valvate, about $^1/_6$ inch long; glabrous; a few conical hairs lower down the tube. *Stamens*: 5, epipetalous in the throat of the corolla tube, alternating with the corolla segments. *Filaments*: Scarcely any, or exceedingly short, inserted over the bottom of the division of the corolla. *Anthers*: Oblong, glabrous, half within the tube, half out. *Style*: Of the length of the corolla tube; glabrous; filiform. *Stigma*: Small or short; undivided, capitate; sometimes indistinctly 2-lobed. *Ovary*: Free; 2-celled. *Placentas*: Fleshy; adnate to both sides of the dissepiment. *Fruit*: A berry, globose, smooth, indehiscent, with a fragile shell-like pulp, which is intensely bitter. *Seeds*: immersed in the pulp, 2-5 in number; ½ inch in diameter, circular, discoid, shining, light grey, silky; not reniform as Brandis says, but having one surface convex, and the opposite correspondingly concave, with a small foveola in the centre of each side. *Albumen*: White, horny or cartilaginous as Gaertener calls it. *Embryo*: Very small, compared with the size of the seed; straight, eccentric; milk-white. *Cotyledons*: Cordate, acuminate, tri-nerved, very thin. *Radicle*: Clavate, very small, placed near the hilum.

The wood is very hard and close-grained, white or grey, with numerous medullary rays. One cubic foot weighs 52 pounds. It is used for many purposes, such as ploughs, cart-wheels, cots, and fancy cabinet work.

Lieut Col K R Kirtikar was born at Mumbai in 1844. After medical training in the Grant Medical College he went to England to compete for the Indian Medical Service. Soon after his return to India he was sent out from 1878 to 1880 on field service in the Second Afghan War where he distinguished himself for gallantry at the Battle of Maiwand. In 1902 he became Brigade Surgeon Lieut Col and retired from service in 1904 to take up permanent residence at Andheri (now in Greater Mumbai). One subject in which he had shown special interest and marked talent throughout his career was botany. There is no department in botany, except perhaps Physiology, which he did not cultivate.

A posthumous work of his on Indian Medicinal Plants was published in 1918, and revised in 1933, and is a standard manual on the subject.

The tree appears to be a native of Ceylon. My description is mainly drawn from the specimens obtained from the two handsome trees growing at Bassein in the Salsette Island in a garden near the ruins of the old Portuguese Fort.

Remarks

Every part of the plant is exceedingly bitter, particularly the root. The pulp of the fruit, says Roxburgh, "seems perfectly innocent, as it is eaten greedily by many sorts of birds." Colonel Drury quotes this observation in his "Unusual Plants of India".

The root has the reputation of curing intermittent fevers. Rheede says that when boiled and drunk, it is purgative. The bark is used as an antidote for snake-bite. Brandis says that the pulp in the fruit is orange-coloured. It is not so; it is white. It is difficult to understand how such a careful observer as Brandis says so. It is evidently a misprint or slip of the pen. The seeds contain 0.28 to 0.50 per cent, of an alkaloid called *Strychnia*, mixed with another alkaloid *Brucia*, closely related to it. *Igasuric acid,* similar to *malic acid,* is associated with these alkaloids. It is these alkaloids which render the plant poisonous.

The late Professor Sir Robert Christison says that the bark might be advantageously substituted for the seed in the preparation of strychnia.

The tree flowers in the cold season. Kurz in his "Forest Flora of British Burma," (Vol. II, pp. 66-167), says it flowers in April and May. It may be so in Burma. The trees in Bassein flower in January. The fruit is ready in the early part of the cold season. Kurz says that the tree sheds leaves in the hot season. It is not known to do so in Salsette.

Brandis says the seeds are flat. If it be so, it is quite exceptional. The general form of the seed is correctly described by Gaertener when he calls it convexo-concave.

Roxburgh observes in his "Coromandel Plants" that the shell covering the fruit is somewhat hard. It is not so when mature and dry. It has the appearance of being so when the fruit is but half developed and the pulp has not yet become jelly-like, but is dense and comparatively drier. When, however, the fruit matures and the pulp is well formed and becomes almost isolated from the shell, the thinness is apparent. It is still more so when the fruit becomes dry; the seed and the pulp then lie loose in the cavity, and the shell easily cracks with a resinous fracture when pressed between the fingers.

Poisonous Properties

Nux-Vomica is so well-known for its poisonous properties that it is hardly necessary to do more in these pages than state them briefly.

Strychnine tree *Strychnos nux-vomica* (Linn)

Strychnos nux-vomica is a common tree in Sriharikota

Strychnia, the chief active principle of this plant, is one of the most powerful poisons acting on the nervous and muscular systems. It causes tetanus — that is to say, tonel contractions of all voluntary muscles. These contractions are generally sudden and last from a few seconds to many minutes. They follow each other in rapid succession. In severe forms there is hardly any intermission. The whole body in such cases becomes "rigid, immovable, and hard as a board" (Schmiedeberg).

The convulsions excited by this alkaloid originate in the spinal cord probably by acting directly upon the motor-cells. The reflex irritability of the spinal cord, of the medulla oblongata, and of the brain is excessively increased. This causes tetanus. When the brain and medulla oblongata are in this state, the spasms get excited by the slightest, often imperceptible, stimuli, which may meet the eye, the ear, and particularly the organs of touch, so that they apparently come on without a cause (Schmiedeberg).

Strychnia has been found in blood. It has a marked effect on circulation. The blood pressure rises; there is arterial tension during the appearance of the convulsions; the frequency of the pulse becomes simultaneously slowed. This, Hayer believes, to be due to vaso-motor spasm from increased irritability of the origins of the vascular nerves and the cardiac inhibitory fibres of the vagus.

It must be remembered that the mind is perfectly clear in strychnia-poisoning. Strychnia is a cumulative poison. It also diminishes the process of oxidation in blood — that is to say, the amount of oxygen absorbed and of carbonic acid given out by blood are diminished (Harley).

Brucia is another alkaloid found in Nux-Vomica, but in smaller quantity than strychnia. It possesses properties similar to strychnia, but as a poison brucia is less active than strychnia.

■ ■ ■

The Kathiawar Lion

By Lieut Col L L Fenton (Indian Army)
Year of publication: 1909

Lieut Col Laynard Livingstone Fenton, was born in 1849, and was commissioned to the 1st Grenadiers, Bombay Native Infantry (101st Grenadiers, IA), in 1872. He joined the Royal Artillery in 1879. In the following year he joined the Indian Revenue Survey, and was appointed Political Assistant at Kathiawar in 1889, and President of the Rajastanik Court in 1896, which appointment he held until he retired in 1901. During the War he offered his services to the Government of Bombay. He passed away in 1921.

Col. Fenton was well known in many parts of India as a keen sportsman and naturalist and contributed frequently to various sporting journals. The preservation of the Lion from extinction in this country is due very largely to his efforts.

Col. Fenton is to be classed among those intelligent Field Naturalists who by their carefully collected notes and observations do so much to assist the Scientific or Museum worker in unravelling the many problems before him.

In spite of the fact that a certain amount of protection is accorded to Gir lions by the Junagadh Darbar, there cannot be slightest doubt that they are gradually, but surely, approaching extinction. Not so very many years ago they were to be found in fairly considerable numbers in the country round Gwalior, Goona, Saugor, Khandesh, Jhansi, and even as far eastward as Allahabad. The districts, around Mount Abu, Deesa and Ahmedabad, along the banks of the Sabarmati river as far as the Rann, were also favourite localities for them. In an old sporting magazine I have read that in the year 1832 the officers of the 23rd Bombay Cavalry used to hunt lions on horseback in the Deesa district, in what way it was not stated, and an old well-known officer, formerly of the Central India Horse, informed me that during the time he was with this regiment no less than 26 lions were shot by the officers in Central India.

MEETHIL MOMAYA

They have, however, long since disappeared from all these localities. The last lion that was, I believe, shot outside Kathiawar, was shot on the Deesa race-course, by the late Colonel Heyland of the old 1st Bombay Cavalry. This was over 40 years ago. It was rumoured, a few days ago, that another one had been seen somewhere in the same neighbourhood, but this could not be proved. In Kathiawar itself, some lingered for a time in the Barda and Aleche Hills in the South and in the wild tracts round Chotila, known as the "Tanga", and in parts of Dhrangadhra, Jasdan, and a few other States in the North of the Province. Then they were heard of only in the Gir Jungle which has always been their home in the Girnar Hill, which, before it was isolated by the march of cultivation, was practically part and parcel of the Gir, and in the Barda Hills, which lie about 10 miles north of the Port of Porbandar, a very rugged group measuring about 10 miles across, covered where the soil allows of it, with low jungle, which also before their isolation owing to the same cause, were connected with the Gir by way of the Aleche hills and the then rough country extending between Dhank and Chorwar on the sea coast. When, however, with the gradual settlement of the country, these last two favourite haunts were cut off from the Gir by cultivation, the lions were compelled to desert them too, and

MEETHIL MOMAYA

confine themselves to the Gir. The story goes that the Bardas were deserted by them in consequence of the guns fired on the hills by the British Force sent in pursuit of the Waghir rebels. Doubtless they disappeared about the same time, but I am confident that the *real* reason for their doing so is the one I have *stated* above. At one time, they must have been fairly numerous in the latter hills, which before the famine abounded in their natural food, viz., sambar and pig besides being the grazing ground in the hot weather of all the cattle in the low country surrounding them. The late Jam Vibhaji of Navanagar told me he had shot lions there as a young man, and there is a curious fresco painting on the walls of one of the rooms in the Lakola at Jamnagar depicting a former ruler, viz., Jam Ranmalji, engaged in the same sport, in the company of his Bhayat with a following of Khawases and armed retainers.

Occasionally, even now, during the monsoon when the crops are high, a lion or a party of them find their way into the Bardas as well as into the Girnar. At the commencement of the Porbandar Administration, about 23 years ago, a party of three, viz., a lion, a lioness and a cub, made their appearance in the hills. Mr. Sealy the then Administrator, wished to preserve them, but they were done to death by the Rabaris and the Navanagar Police stationed there at the time, to keep out the Mekrani outlaws against Junagadh. I saw the skin of the lion afterwards in the possession of an officer: it was a very fine animal with a fairly good mane. The Girnar Hill being so much nearer the Gir than the Bardas, occasional visitors to it are not so rare; I was told by Mahomed Khan, the successor of the old Balooch Inamdar of Kadia, a village at the base of the south-eastern slopes of the hill, that a few years ago a young lion made its appearance in his village and killed a cow belonging to one of the villagers. It was followed up the next morning by the owner of the cow, a Mekrani sepoy in the service of the old Baloochi. He came suddenly upon the lion in the act of devouring the carcase, on the out-skirts of the village, in a prickly-pear thicket. The lion charged at once, knocking over the sepoy and mauling him badly, but the latter kept his presence of mind, and succeeded in driving the beast off, after inflicting such severe wounds upon it with a "jambia" or short covered dagger, that it succumbed to them before going any great distance. The old sepoy, when I saw him, not long after the encounter, had quite recovered from his wounds.

Another adventure with a stray lion took place in a Sindi village, not very far away from the same neighbourhood. The story was told me by an eye-witness. In this case, a cultivator, early one morning while on his way to his fields, came across a lion devouring a cow it had just killed. He immediately hurried back and gave the information in the village when practically the whole of the village population turned out armed with tom toms, empty tins, lathis, etc., for the purpose of driving the unwelcome visitor away. On seeing the crowd approaching, the lion left the "kill" and retired into some bushes, whence it declined to stir, in spite of all the efforts of the villagers to make it do so. Some of the men bolder than the rest managed to reach and climb into some trees overlooking the bushes into which the lion had retreated and tried to make it move by pelting it with stones, but all to no purpose, a few ominous growls was all they elicited in response to their fusillade. At this stage in the proceedings, a "Rabari" appeared upon the scene, a cattle-herdsman by caste and profession, and a member of one of the handsomest, pluckiest and finest class, of the many to be found in the Province. On learning from the villagers the cause of all the uproar, he, instead of following their example of joining in the fun from the same position in a tree, laughed at them for their cowardice, and declared he would single-handed very soon put the lion to flight. To put his boast into effect, he at once proceeded to walk towards the spot where the lion was said to be crouching, shouting at the top of his voice and brandishing his lathi as he did so, and doubtless quite convinced in his own mind that the lion would turn tail and bolt on seeing him steadily approaching, and probably it might have done this, under any other circumstances, but it is not surprising that, after all the baiting that it had undergone at the hands of the villagers, its temper at the time was not of the sweetest, and that instead of at once decamping, it charged and laid low the unfortunate "Rabari" with a gaping wound in his side which rapidly proved fatal.

MEETHIL MOMAYA

A short account of the Gir Lions, in whose haunts I have lived for weeks together and with whose habits I have therefore perhaps had better opportunity of becoming acquainted than most members of our Society, may not be out of place in our journal, especially as I do not remember to have ever before seen them mentioned in its pages. The Gir forest, where only the lions are now found, covers an area of about 1,500 square miles within the territories of the Nawab of Junagadh. The greater part of it is covered with a jungle of stunted trees composed principally of dwarf teak, jambool, khizda, khakra, kadaya, bor and babul with here and there, patches of bamboos, corinda and other thorny bushes, and an isolated wadh or banyan tree towering far above its neighbours. The country is undulating with a few rugged hills in parts and much cut up by nalas, with rough rocky beds, and their banks as often as not lined with a thick growth of jambool trees. The "Thran" river is the largest stream in the forest, and in ordinary years it, with some of the larger nalas, holds water all the year round in the deeper pools. Rock is almost everywhere near the surface, which accounts for the stunted growth of the trees and, I imagine, the rank coarse vegetation which covers the jungle during the monsoon, rendering it almost impenetrable in parts, at that season of the year. Villages, if a collection of dilapidated huts surrounded by patches of cultivation can be looked upon as such, are few and far between. Sasan, which is the headquarters of the local Darbari official and where shooting camps are as a rule pitched, may be looked upon as the capital of the Gir. Nesses or hamlets, being collections of temporary huts, the dwelling places of local herdsmen such as Rabaris, etc., are scattered in suitable localities all over the forest. As might be expected a very bad type of malarious fever prevails both in the Gir and in parts of the Girnar. The greatest sufferers from it are of course the outside cattle graziers who visit the forest only at certain seasons of the year, with swarms of cattle for temporary grazing purposes, and more specially so the cultivators and their families, who are from time to time imported by the Darbar into the Gir whenever it is considered advisable to establish a new village. In the village of Hasnapur, comparatively recently established in the crater of the Girnar, almost every soul I saw was suffering from enlargement of the spleen. A former village on the same site had undoubtedly been wiped out by the same disease. The site is admirably adapted

for a game preserve and I am surprised the Darbar does not reserve it for such a purpose, instead of keeping up the village. The panthers have only to be kept down; Sambar, Pig, *Gimtada* (Four-horned Antelope) are already there, and Chital might be imported as a trial. The actual natives of the Gir are, as might have been expected, practically immune from the fever. The most noticeable class amongst these are the descendants of men who were originally imported into the country from Africa by the Darbar probably to serve as mercenaries, and who intermarried with the natives and settled down in it for good. Many of them are still to be found in the service of the Darbar in the ranks of the police sibandi, etc. Physically they are a very fine set of men, and some of the best shikaris and trackers in the world. They are the shikaris of the Gir, and no lion shoot is ever undertaken without the services of certain well known men amongst them being called into requisition. For many years one Hebat of Jambuda was considered the best man and took the principal part in all the big shoots—but of other younger men, coming on in the same direction there is no scarcity. So much for the lion country and its people. As regards the wild beasts to be found in the Gir besides the lion, as far as I am aware it was never the resort of the tiger or the bear and it may be added by the way that it does not hold any description of jungle or spur fowls. Of the undermentioned animals however when I knew the Gir, in its prime, before the last famine, the jungle was practically full, viz:-

	Local vernacular name
Panther	*Dipdo.*
Hyena	*Jarak.*
Pig	*Soor* or *Kalajanwar.*
Sambur	*Sembar.*
Spotted Deer	*Pasu.*
Four-horned Antelope	*Gimtada.*
Nilgai	*Roz.*
Gazelle (in the more open parts)	*Chinkara.*
Blackbuck (on the outskirts)	*Kalyar* (doe, *reda*).

The local vernacular name of the lion is *Sawaz*, i.e., one who causes the flocks to bleat. Sometimes but very rarely it is called the *Untia vagh* obviously from the fact of its colour being somewhat similar to that of a camel.

In the matter of food, therefore, the lions were well off with the game alone, in addition, they had the swarms of cattle which were brought into the Gir from outside to graze and which undoubtedly paid a heavier toll to them than did the wild game.

I have not paid a visit to the Gir since the last famine, but have been told by others who have done so, that in the matter of game, it is a very different place to what it used to be "in the good old days". Until towards the end of the famine by which time the Gir had been pretty well cleared of both wild game and cattle, the lions and panthers fared no worse than usual, but it was a very different state of things for all the deer kind: not only did their natural food very soon fail them but from the very commencement of the famine they were

MEETHIL MOMAYA

mercilessly persecuted by the local Darbari police, forest guards, etc., who were able to and did shoot down hundreds of animals over the puddles of water left in the district, in spite of the orders of the Darbar prohibiting their slaughter. In the depths of the Gir the lower Darbari officials do pretty much as they like. An old sepoy once laughingly said to me that he was accustomed to eat meat and the Gir was the only place where he could get as much as he required. This was when the game was supposed to be preserved!

Towards the end of the famine the lions as well as the panthers began to find their food was running short, they were therefore forced to leave their usual haunts, and wander in search of it into the surrounding districts. This brought them more into evidence, and gave rise especially at the time of Lord Lamington's shoot which terminated so disastrously, to the rumour that owing to the very strict protection which had been accorded them, the lions had increased enormously in numbers. This I feel safe in stating was not the case. As a matter of fact the preservation was never very strict. The Darbar was always very liberal in granting the local officers and others permission to shoot a lion; all the cubs captured in the Gir were invariably sent to Junagadh to be placed in confinement in the gardens, and I was told as a fact that the Rabaris or local cattle herdsmen, who naturally had no love for the lions, made away with any cubs they came across if they found they could do so without fear of detection. Moreover, although there are no lions in Baroda territory, which bounds the Junagadh Gir on the east, some of the best jungles for lions on the Junagadh side abut on this boundary and I should be sorry to say how many lions have been killed by their not keeping within their own limits, to put it as mildly as possible. With no recent information to go upon I cannot give an approximate estimate even, of the number of lions in the Gir at the present day, but it may be taken for granted that if the Gir is allowed to be cut down in the future as it has been in the past, the day is not far distant when the Indian lion will have become extinct. Fortunately the loss of the forest would mean far more, not only to Junagadh itself but also to the whole of Kathiawar, than the loss of the lions, so let us hope that they will be spared for many a long day yet. Of course, the lions do a fearful amount of damage among the cattle, but this might be remedied to a great extent by properly preserving the game animals which are their natural food. These are, it is true, *nominally* preserved now but as a matter of fact the forest guards and police, who are supposed to be the gamekeepers, have an understanding amongst themselves that the game laws are intended for others and not for themselves. For the better preservation of the lions the sooner the wholesale and indiscriminate slaughter on the part of these subordinate officials is stopped the better. I have often wondered why the Darbar does not close some 600 square miles of the Gir not only as a reserved forest on the lines obtaining in British India, but to serve also as a sanctuary for the lions and all descriptions of wild game. It is worth the trial and there are several localities well adapted for the purpose, notably the country including the Nesses of Sirwan, Khokra, Chelna, Moduka and Jamwadla.

It is curious that the old idea, that the Indian lion is a, maneless one, still prevails amongst a host of people not excepting sportsmen who have never had an opportunity of seeing the animal. Anyone who has taken the slightest interest in the subject is of course well aware that such is not the case. It is true that in a wild state the Gir lion does not carry as heavy a mane as the African, but this comes of the former's home being in a thorny jungle where its mane is bound to suffer, whereas the latter is more or less a dweller of the plains. In captivity there is not much to choose between the two in this respect, although I have noticed that in the Indian animal the mane does not extend so far under the body as it does in the African. It has also been stated that the Indian lion is a much smaller beast than its African brother. To decide this question we can only refer to the measurements taken and recorded by sportsmen, and unfortunately very few measurements of the Indian animal are to be found recorded anywhere. Moreover, there is nothing to show that in every instance the measurements were taken in exactly the same way, which must have been done for them to be of the slightest value for purposes of comparison. It was stated shortly after Lord Lamington's shoot already referred to, that of the lions shot by his party one, if not two, measured over 11'! Subsequently it transpired that the measurements were taken *after* the animal or animals had been skinned. I was unable to ascertain whether the measurements stated recorded the length of the skins or of the bodies after the former had been removed, anyhow the measurements are obviously of no value, and it may be regarded as a certainty that a lion of the dimensions stated—the measurements being taken in the recognized way before the removal of the skin—never existed in the Gir or anywhere else in India. of four Gir lions shot, and very carefully measured by myself, the total length of the largest was 9'-5", the length of the tail being 2'-1". Two of the others measured, respectively, 9'-1", and 9', both being younger animals than the first. The one shot by Lord Harris measured 9'-7", another by the late Lieut. Percy Hancock was a still finer beast but unfortunately its measurements were not taken.

In Rowland Ward's Book of Measurements, 3rd edition, mention is made of African lions measuring from 9'-1" to 10'-5" only, four being 10' in length and some ten below 9'-5" not including lionesses of course, as they are always smaller than the males. In the same book of records, I notice it is recorded, that a 10'-4" lion has a body measurement of 7'-2", a 10" lion, of 6'-10' and one of 9'-8", a measurement of 6'-6½"—this goes to show that the tails vary considerably in length and the weight of an animal cannot be judged correctly simply from its total length. Besides, abnormally large sized specimens are to be found in every description of animal life. Moreover, when comparing the whole area of Africa with a small one like that of the Gir Jungle— "a drop in the ocean" — it is by no means extraordinary that the few larger specimens should have been recorded from Africa, especially as so few measurements of the Indian animal are forthcoming. For a true comparison we must look to the *average* measurements, and these undoubtedly prove the latter to be every whit as fine a beast as the African.

A comparison of the skull measurements is interesting. Those of my 9'-5" lion are as follows:

Total length between uprights	13.4 inches
Width across the zygomatic arches	8.6 inches
Height resting on table	6.2 inches

In Mr. Rowland Ward's Records of Measurements, those of over 30 lions run much higher than the above—the largest measuring no less than 16'-5" in length. But these, of course, belong to *picked* heads from all parts of Africa, and they certainly do not shake my belief that taking the *average* there is no difference in point of size between the two animals. A lion, which was presented to the London Zoological Gardens by the late Colonel Humfrey, was as fine a specimen as any of the African lions in the adjoining cages, and the lion confined in the Sardar Bagh at Junagadh will compare favourably in the same direction with any African specimens in confinement in any part of the world.

My 9'-5" lion's skull measurements compare as follows with those of a very heavy old 9'-8" tiger I shot in North Korea.

	Length.	Skull Length.	Breadth.	Height.
Tiger	9'-8"	13.7 ins.	9.3 ins.	6.3 ins.
Lion	9'-5"	13.4 ins.	8.6 ins.	6.2 ins.

The principal difference is apparent in the *breadth,* the tiger's being consequently the much heavier looking skull.

The chief difference between the skulls of the two animals lies in the nasal bones, the posterior terminations of which, in the lion, are opposite the terminations of the maxillary bones, whereas in a tiger, they extend beyond them. The lower part of a lion's underjaw is also convex and does not sit flat on a table like a tiger's does.

In a description given of the Indian lion by the great authority, Lydekker, in his book on the Great and Small Game of India, Burma and Tibet, he considers it possible that a claim to racial distinction between the Indian and African animals may be drawn from the colour of the mane. He states that he himself has never *heard* of the occurrence of a black maned lion from the former country and further mentions that it is definitely recorded by a Colonel Percy in the Badminton Library, that black maned lions are *absolutely unknown* in India. I cannot, of course, say what grounds Colonel Percy had for making such a downright assertion applicable to the whole of India. He evidently had no experience of the Kathiawar lion, for there is sufficient evidence, I consider, satisfactorily to prove that black maned lions have been known to occur in that Province.

The evidence I refer to is as follows:

(1) Many of the Gir *pagis* including one old and very celebrated one of the name of Hebat, who, I fancy, is now dead, have *over and over* again, told me that black maned lions did occasionally occur in the Gir and had actually been seen either by themselves or their fathers before them. Those men were not called upon to settle the question of manes—they only knew the lion of their own country, lions with black manes had been known to exist, and therefore they stated as such. By the side of further evidence I had no reason to doubt their word, although I had not myself come across a lion with a wholly black mane.

Photographed in captivity at the Junagadh Zoo
(Photo: C.H. Hill I.C.S.)

(2) The late Colonels Watson and Scott, both of whom were very well acquainted with the Gir and its lions—the former especially so—have on more than one occasion mentioned in my presence the rare occurrence of lions with black manes in the Gir. The former was a very observant officer and a great *shikari,* and one who would not have made such a statement had the slightest doubt existed in his mind on the point.

(3) In an old Agency document which came before me in a case in which the Junagadh Darbar and an old Kathi Chief, one Harsur Khachar (formerly of Chelna in the Gir) were the interested parties, (I am stating these particulars in order to locate the evidence should any one hereafter care to see it), it is incidentally stated to the effect that "Colonel Le Grand Jacob, while on his way to the Gir to shoot *a black maned lion,* had been obliged to give up the expedition and return to Rajkot to transact some important business which had to be seen to without delay."

MEETHIL MOMAYA

This piece of evidence is in itself sufficient, I consider, to remove all doubts upon the point.

(4) I saw the lion mentioned before which was shot by the late Lieut. Percy Hancock. It was a fine beast with a good mane. *I noticed several black locks in the latter.* Of course, a few black hairs do not make a black mane, but the black locks were unusual, and had the lion lived, the whole mane might in course of time have turned black.

Shortly—the occurrence of black manes amongst the Kathiawar lions is extremely rare, but that they do occasionally occur I consider there is sufficient evidence to show. More evidence on the point could possibly be adduced by any resident interested in the subject, from the Gir *pagis* who would certainly have noticed the occurrence of any black manes amongst the lions of more recent generation.

The lion is a far more noisy animal than the tiger and for this reason is more easily brought to bag, being so much more in evidence. They generally commence roaring early in the night and often keep it up until the dawn for no apparent reason. When arranging for a shoot it is usual, besides tying out kills, to locate *pagis*, in parties of two, in different parts of the jungle overnight; these ascertain during the night the direction in which any lions are moving, by their roars, and as soon as it is light enough to see, pick up the trails which they never leave until they have marked down the animal or animals, as the case may be, either busy over a kill or more often resting for the day, resting under the shade of a large *wadh* or banyan tree, or sometimes, in a *bohira* or waterhole to which they are very partial. A *pagi* told me that while watching one evening over one of these *bohiras* he saw a hyena, two porcupines and finally two lions emerge from it at short intervals! On another occasion, while watching over one of them myself' no less than three lions came out of it. The first one that did so was shot by one officer who was watching with me, but when the remaining two came out, it was too dark to see them, although I heard them growling only a few yards away under my tree. Since the lions, as a rule, cover a great deal of ground during their nightly wanderings, the spot where they are eventually marked down may be many miles away from camp.

To show the amount of damage the lions do amongst cattle I may mention that the last one I shot was one of two which had during the previous night broken into the cattle zareba of a Rabari Ness, killed three cows and mauled two more. A sixth was missing when I arrived on the scene. This one we found later on quietly grazing near the lions while they lay fast asleep under a *wadh* tree quite two miles away from the Ness, having evidently been driven along by them to serve for their next meal.

"We do not know why Emperor Asoka installed the Lion atop the pillars bearing his edicts, in 3 BC. Maybe he chanced upon a male Lion standing on a mound, watching over a vast plain stretching to the far distant horizon, conveying a rare picture of sovereign self-assurance, feudal overlordship and pride with dignity; in essence, the King of all he surveyed!

Over time, the stylised Lion of Asoka's pillars, also became the symbol of the Sovereign Socialist Republic of India in 1952. Thus it has become a tacit article of faith that the Asiatic Lion shall have a permanent home in India for all times to come."

Lieut - General Baljit Singh

■ ■ ■

A History of Shikar in India

By Lieut Col R W Burton, (Indian Army)

Year of publication: 1952

The India of our subject includes the whole subcontinent, also Burma and Ceylon. We have to pass in review the Indus Valley flanked by the Kirthar, Baltistan and Suleiman Ranges and then see Kashmir and adjacent territories of Baltistan. Ladak and Changchenmo, Zaskar, Rupshu, Spiti and Lahul all of which are a vast entourage of snowy mountains, riven ravines and precipices; of plateaux and lofty ranges which remain an everlasting wall between India and the rest of Asia.

"Northwards soared the stainless ramps of huge Himalaya's wall"

Where the mountains have a northern aspect they are usually forest covered, while the southern slopes and folds of the hills are often bare and dry, subject to forest fires and the depredations of domestic flocks and herds.

Lower grew the rose oaks and the great fir groves where echoed pheasant's call and panther's cry

Continuing east we pass over the wooded and often mountainous tracts of the Simla Hill States, Garhwal and Kumaon until we meet the five hundred mile long exclusive Kingdom of Nepal. Then we see Sikkim and the dense forests of Bhutan, which have been almost unknown to sportsmen of the past and present alike, until we arrive at the northern part of Assam, so often devastated by earthquakes. Here we may remark that the animals of the Eastern Himalayas resemble those of the Burma region, while along the mountains to the westwards are kinds more akin to those inhabiting the temperate parts of Asia. Passing over Burma, Tenasserim and the Malay Peninsula we view Java and Sumatra and then turn west for Ceylon. Within that enormous arc is the Peninsular India with which our subject largely deals.

What is Sport?

It can be said that all sport is governed by unwritten laws, and the general tendency is to give the animal a sporting chance of escape also to make the sport as great a test as possible consistent with the object in view—the death of the quarry. It may also be defined as measured by difficulty in achieving success.

Born in 1868, **Richard Burton**, the sixth son of Gen. E.F. Burton of the Madras Staff Corps, was commissioned from Sandhurst in 1889. Posted to the Indian Army in 1890, he was permanently crippled by a riding accident in 1903. He was, thereafter, assigned to the Cantonment Magistrates Department. A fearless sportsman and a keen fisherman, he wrote over 200 articles on variuos aspects of Natural History and was the first Naturalist to campaign for the preservation of Indian Wildlife. Col. Burton passed away at his residence at Surrey, England in January 1963 at the age of 95.

The Pre-Mogul Period

The physical aspects of the Indus valley have undergone many changes. No longer are there the forests which provided timber for the first Indus flotilla constructed by Alexander in 325 B.C.; gone are the rhinoceros and the elephant; gone are the swamp deer, and the last tiger was shot in 1886. Hog-deer, wolves, chinkara, wild dogs, jackals, hares, cats, and the hyena very rarely, now comprise the larger animals of the Indus valley. The Indian Antelope (Blackbuck) has been introduced into the Khairpur territory.

In the early Jain and Buddhist periods (*c.* 600 B.C.) there was considerable knowledge of mammals, birds and reptiles, but previous to the appearance of the Emperor Babur on the scene there is little information concerning shikar.

When the Moguls first entered India in 1526, the Rhinoceros was along the Indus

The Mogul Period

From 1526 to 1707 much of interest is contained in the memoirs of the Mogul Emperors and the chronicles of European travellers in India in those times. The famous illustrated copy of the *Ain-i-Akbari*, bearing the signature of the Emperor Jehangir, in the Victoria and Albert Museum should be seen by all who can do so. In the series by Sálim A. Ali on "The Mogul Emperors as Sportsmen and Naturalists" we learn about the hunting methods practised in those days; and this is aided by Handley's valuable illustrated article. These two contributions afford a remarkably full picture of the shikar methods and natural history knowledge of the period. The shikar grounds of the Moguls were the upper valley of the Indus towards Peshawar, and the whole of the present U.P. westward from the Ganges to Kathiawar and southwards to Mandla in the Central Provinces.

The Shikar Animals of the Moguls

Elephant, rhinoceros, buffalo, were known to the Moguls, but not the 'Bison'. When the Moguls first entered India in 1526, the rhinoceros was along the Indus, and the elephant in many places whence it has since vanished. Akbar was specially interested in trapping wild elephants. At the present time there are no longer any elephants north of the Dehra Dun Siwaliks; the rhinoceros lives only in Nepal, Bengal and Assam; the wild buffalo in those same areas, while a few herds survive here and there in Orissa, Raipur, Jeypore and Bastar.

The Larger Felines: The Emperor Babur was a fine sportsman, as also was Akbar, while Jehangir excelled as a naturalist. Akbar disliked the less hazardous methods of tiger-hunting—traps, nets, limed leaves, etc.—and preferred to attack these animals openly with bows and matchlocks.

In Mogul days, and as late as the 1830s lions were numerous in Hindustan. Jehangir killed them in Malwa, and the Rev. Terry (*c*. 1650) was frequently terrified by them when passing through the then vast jungles of that country.

The Mogul Emperors quickly discovered the delights of Kashmir, but there is little record of what they did there in the way of shikar. Abul Fazl mentions that the snow leopard was tracked in the snow in Kashmir, but since this is a very elusive animal, seldom seen by sportsmen, it is more likely that this had reference to the common leopard or panther. This is still considered a fine sport by the few who have done it.

Bears: Of bears there seems to be almost no mention in the Mogul literature.

Deer: Nor do we find much about hunting of deer in the Mogul days. A net was put round the horns of a tamed deer and the horns of the wild one became entangled. It is related that one of the deer "caught" a leopard which became entangled in the net. The species of deer referred to is not clear. Another form of hunting was by means of a light inside a basket on a man's head; the animals attracted were shot or speared. The modern poacher uses electric torches or other contrivances and buckshot cartridges.

Antelope and Gazelle: There must have been a very great number of antelope (blackbuck), nilgai and gazelle in all the areas suited to them. All the Emperors, Jehangir in particular, were extremely fond of hunting the nilgai and spared no personal effort in pursuit of sport where this species was concerned. Blackbuck were trained as decoys to take the wild ones by the net method. That same device is in use in a part of South India at the present time.

Hunting with the Cheetah: This is a pastime indulged in by many notables in India since very early days. The Mogul Emperors were partial to the sport, and Akbar kept a thousand of these animals. Three sets were khâcâh ("Royal") or for use of the sovereign. The monarch's, best leopard, by name *Samand malik* ("like a ruby"), rode in a *chandol*, or litter borne on the necks of two horses.

In a wild state the cheetah hunts antelope, gazelle and the smaller deer, also hares, peafowl and other birds and the smaller mammals, but for sport it is mostly trained for blackbuck. The buck is struck down at full speed, not by blow of a paw only as is commonly stated, but by use of the large-taloned dew-claw which gives the necessary purchase. Blackbuck can attain a speed of 42 miles per hour when hunted and going all out. The cheetah is an animal partial to rocky and open country and was soon shot out when the land became more developed. They were frequently found in packs, and there is record of a cavalry officer having in one day speared six off one horse. The animal having become exceedingly scarce in India, the support for sporting purposes comes—or used to come latterly—from Africa. The animals have to be trapped when full grown; if taken as cubs the training is tedious and unsatisfactory.

The Caracal—"Siah-gôsh" as the Moguls knew it—is easily tamed, and was trained in the same way to kill gazelle and the smaller deer, foxes, hares, peafowl. Vigne witnessed the sport and says their speed is, if possible, greater in proportion even than that of the cheetah.

Falconry: The antiquity of falconry is known to be very great, and it is certain that the Moguls gave much impetus to the sport in Northern India. In the Sálim Ali series we have something, also in Handley's "Sport in Indian Art", where we learn that Akbar hunted with trained falcons and hawks of which his favourite was the bashah (Sparrowhawk). In the *Ain-i-Akbari* names of many varieties are given, and the names of those in use in Sind are in Langley's book.

The famous French physician, Bernier, relates of the Emperor Aurangzeb that there passed before him at his daily Court, or Public Audience, "… every species of the birds of prey used in field sports for catching partridges, cranes, hares, and even it is said for hunting antelopes, on which they pounce with violence, beating their heads and blinding them with their wings and claws."

Fishing: The Mogul Emperors were partial to the ancient sport of fishing. It is common knowledge that Muhammadans of the present day all over India are much addicted to angling with rod and line in both rivers and lakes; and there are many anglers in Bengal and other parts of India also.

The Post-Mogul Period

Tiger and Lion: Judging by the number of tigers and other game in a seventy by thirty-mile area near Neemuch in the years 1850-1854 as related by Rice, and the mention by Newall of a railway official having killed one hundred tigers in Rajputana owing to the facility with which he could move about, the quantity of game in the time of the Moguls must have been very great. Gordon Cumming takes the modern record to the Tapti river border (in 1862 ten tigers in 5 days); Montague Gerard killed 227 tigers in Central India and Hyderabad before he left in 1903; Prideaux of the Central Provinces shot 147 tigers during his service up to about 1930. Forsyth, Hicks, Glasfurd, Burton and others fill in the period 1845 to 1905 as to the land of hills and plains from the Narmada to the Kistna. For Madras (Tamil Nadu) and Ceylon there are Campbell, Hamilton, Sanderson, Samuel Baker, Dawson, Drury, Fletcher and some more.

In regard to Orissa, Bengal, Assam, and Bihar to the Siwaliks we have Williamson, Okeden, Kinloch, Simson, F.W. Pollok, E.B. Baker, Fayrer, Baldwin, Braddon, MacIntyre, Adams, Lambert, and others to fill in the hundred years from 1780 to about 1880.

In 1852 a tiger killed an officer of the 98th Regiment 23 miles from Rawalpindi; there was a man-eating tiger near Poona in 1849; and there are interesting records of tigers on the islands of Bombay and Salsette. Owing to increase of cultivation and decrease of forest, tigers are in less number than formerly. Although people are still killed by them in some tracts they are necessary to the forest economy, as are the deer and wild pig on which they are meant to exist, so neither the tigers nor their natural prey should be unduly destroyed by man.

It is said in the *Bengal Sporting Magazine* of 1837 that within 23 years of occupation of the country (after the Mahratta Wars) the lions were extinct in the dry and sandy deserts of Haryana. In 1832-33 cavalry officers at Rajkot shot lions from horseback; one of them being 10 ft. 6 in. long with an 18 inch mane. With another gun (Rice?) he killed 14 lions in 10 days in the Gir forest. There are now no lions out of Kathiawar, and the number in the Gir is estimated to be 247.

Panther or Leopard: Panthers are more ubiquitous than the tiger and less affected by the advance of cultivation. In proportion, the animal is more destructive than the tiger, and under favourable circumstances is more deadly as a man-eater being more agile and active, also more silent and more stealthy. He climbs better, jumps better, and stalks better than the tiger, and can conceal himself almost anywhere.

Thomas Vigne was in Kashmir in 1835 and his book would be a useful reference were it not so rare and difficult to obtain. Adams was a naturalist and ornithologist rather than a sportsman. From his book, and from Newall and MacIntyre who were also in Kashimr about 1851-52; it is known there was then much game in those countries. Not long after that the writing was already on the wall. Far too many animals were shot by sportsmen; and the people of the country, then as now, took heavy toll during the winter months.

The conclusion from perusal of all the old sporting books, dealing also with Kashmir and Burma, is that the steady diminution of all the game animals began about 1780 as to Hindustan, 1840 as to the Western Himalayas, later as to Burma, and is now nearing its climax unless it is halted by all the governments.

In Williamson's day there was the sport of riding on the neck of a "koomkee"— a female elephant used as decoy in capturing a male—and throwing a noosed rope round the head of a wild tusker. "This kind of sport," says Williamson, "cannot be classed among the effeminacies of the day!" The hunting by tracking of the rogue elephant was declared by Sanderson to be the greatest of all sports; and is still available from time to time.

Source: "Oriental Field Sports" by Williamson, 1819

A female elephant used as decoy in capturing the male – by throwing a noosed rope round the head of a wild tusker

Crocodiles: Of the two species of crocodile known in India the river crocodile of the burning ghats and other places takes a man when it has a chance, while the estuarine species is a very dangerous reptile. The Indian Gharial is a fish-eater and not feared by man. Concerning crocodiles and the gharial there are more than fifty Miscellaneous Notes in the Society's *Journal*; among which harpooning in tanks, gharial catching in the Indus river, hints on shooting crocodiles, angling for crocodiles, and poisoning of crocodiles! The shooting of these animals in India can be excellent sport and calls for considerable technique and knowledge of the animals. In jungle streams and pools they take considerable toll of wild life. At p. 75, vol. 1 of Langley's book is a visual account by an officer of a tiger being vanquished by a large mugger.

The Sloth Bear and the Malayan Bear: Up to sixty years ago the Sloth Bear was really plentiful all over the forested tracts of India and Assam from the base of the Himalayas to Ceylon. Because of its aggressive habit when chanced upon in the jungle, the hand of man is against it, so it is now almost or quite extinct in places where formerly numerous. Nowhere has it been protected under shooting rules. Many of these bears were speared from horse-back by Colonel Nightingale in the 1860s. This noted sportsman died in the saddle in 1868 while spearing a panther. The Sloth Bear will probably survive through protection in some of the National Parks and Sanctuaries in course of formation, and in its more remote haunts. Naturalists regard the Sloth Bear of Ceylon as a distinct race.

The Malayan Bear of Chittagong and Burma is a smaller edition of the Himalayan Black Bear and merits little mention in this history of shikar as it is seldom met with or hunted.

Wild Dogs and Hyenas: The Indian Wild Dog has an immense range. In earlier days the packs of these animals in forest areas were considerably larger than now. Apart from distemper and other diseases which keep the numbers in check, the fluctuation of the wild dog population must depend considerably upon food supply—mainly deer, pig and other forest animals. Fortunate is it for India that the species does not attack man, and is not habitually destructive to domestic stock. Should it be more and more deprived of its natural food it may, like the tiger, increasingly prey upon the flocks and herds. The hyena of India is not ordinarily greatly inimical to human life. It is here mentioned as occasionally affording sport to the bobbery-pack, or the horseman with his spear.

Sheep and Goats: When the record Sind Wild Goat (52¾ in.) was shot in the Kirthar Range in November 1912, considerable herds were seen; of present stock there is no news. In Baluchistan the Persian Ibex may not have survived the influx of modern rifles; nor will the toothsome Urial have fared better. The Persian Gazelle may have survived in a few places. Of the stock of all these animals in earlier days there is no literature available to the writer, but they probably existed in considerable numbers. The Suleiman Markhor is also an animal of the Baluchistan Hills. Soldier-sportsmen serving trans-Indus used to have fine sport and secure good heads of this race.

In Adams' day ibex were plentiful in Kashmir and Wardhwan; now they are no longer there, and have not been for a number of years. Only in the more remote nullahs of Baltistan, Gilgit and Astor could the sportsman now hope to find worthwhile ibex; and markhor may have almost vanished (Stockley, Vol. 32; 783). Both Adams and MacIntyre pointed out in their book what was happening, and what the result would be; while both Baldwin and MacIntyre remarked on the great diminution of game birds in the Terai and Doon.

Small Game Shooting

In his "Mogul Emperors" series Sálim Ali, being an expert ornithologist, has given us some interesting information. In those days, and up to the period 1840-1860, the game birds of the hills and plains must have been everywhere in great numbers. Nature had evolved for them a high reproduction rate and they were able successfully to contend against all natural checks, and even with the amount of trapping and snaring to which they had been subjected through all the centuries. With the shotgun and its indiscriminate use there came a very great change; but some of this depletion was also due to the increased incentive to the people to snare game for the tables of the foreigners.

Now we have the present intensified diminution of all game birds for there has never been any thought for the morrow, and some species are nearing extinction. During the past few years there has been great opportunity for all game birds to recover in some measure their former abundance, for the changed conditions have made shooting of every description both difficult and expensive. But the apathy of Governments and the authorities, and the activities of trappers and snarers have nullified the opportunity as the demand for meat of any kind has become clamant, and modern communications have made it easy for the supply to reach both markets and consumers.

Failing speedy and suitable measures by Government, the outlook is exceedingly gloomy. Recently, an observer from a foreign land has said to the writer, "You will lose *all* your game birds."

Kashmir: Of Kashmir it is reported at the present time that there is depletion of the number of *chukor.* Large bags of wildfowl used to be made in Kashmir by sportsmen inclined that way. One of these shot 6,998 duck and geese in one year; while another, also shooting alone, bagged 58,613 wildfowl in the seasons 1907-1919. He killed 119 greylag geese in one day, and on another day 509 duck and teal.

Rajputana: In the well-known Bharatpur wildfowl shoots the bags were large. On 20th November 1916 there fell 4,206 birds to 50 guns. Without any reference to anything here written or referred to may be quoted "Some prefer flighting, others shoot for averages and lose many of the delights of an exceptionally high bird, and there are those who will not pull the trigger until three heads are in a straight line!"

In the Imperial Sandgrouse shoots huge were the bags. Perhaps the record may be that of the Bikaner shoot in 1921(?) when Lord Rawlinson was one of the party which killed in two mornings 5,968 birds. Maybe those large bags of wildfowl could still be made, but the world's wildfowl situation does not warrant such slaughter; and perhaps those other big shoots are events of the past not likely to be repeated, for the times have changed.

Pigsticking or Hog-hunting

The sport of chasing the wild boar on horseback with a spear was introduced by British sportsmen in Bengal in the latter part of the eighteenth century. At first the sloth bear was hunted; but in 1776 it was the wild boar, the weapon used in the Dacca District being a short, heavy spear three feet long and well poised. It was thrown like a javelin; and if the sportsman missed his aim he had to dismount and recover his weapon thus letting in the next in succession, and so on till the pig was killed.

The modern spear is up to 6 ft. 3 in. long and fairly heavily leaded, about 1¾ lb. On the Bombay side the spear was eight to ten feet or even more, and often unleaded.

Twenty-five years later a jabbing or thrusting spear was in use in Upper India, but the practice developed in Bengal was to use a spear about seven or more feet in length, also thrown as a javelin as is well described and illustrated in Williamson's ORIENTAL FIELD SPORTS, 1807. We know from Simson that in 1830 the throwing of the spear was discontinued, and penalized by the Calcutta Tent Club at the instance of Mr. Mills, B.C.S. Published in 1880, Simson's book contains complete guidance to everything pertaining to pigsticking in *Eastern Bengal* up to that time; and, except as to localities, is of equal value at the present day.

All regarding the sport as developed in Upper India is contained in the article by Neville-Taylor in the SPORTSMAN'S HANDBOOK FOR INDIA, with which is a map showing the pigsticking centres and Tent Clubs of those parts in 1904.

'Modern Pigsticking' by Wardrop covers practically all of India and is a complete compendium of pigsticking. The Meerut Tent Club country is fully dealt with; also the Kadir Cup which was constituted in 1869 and the winning of which has been the blue ribbon of pigsticking. Among the hazards related by Wardrop is that of a pigsticker's Arab horse, having swum a river, being seized by the head by a crocodile while drinking in shallow water, dragged into deep water and never seen again; and Kinloch, when hunting with the Meerut Tent Club, had his horse ripped, himself thrown and wounded by the boar in *fifty* places!

The Nagpur country is well described by Best and Dunbar Brander.

Praise of the Boar: "It can be said that the finest and most spectacular animal of the Indian jungles is the tiger, and most noble in appearance the elephant; but the concensus of opinion is that the Indian wild boar is the bravest and most gallant of all."… "Nothing for size and ferocity could surpass, if it could equal, the pure Bengali breed; other hunters, however, declare the Deccany pig to be unrivalled for speed and ferocity"; while a widely experienced expert has declared, "Give me a Bengali hog in Guzerat country."… "No man who has not been an eye-witness of the desperate courage of the wild hog would believe in his utter recklessness of life, or in the fierceness that will make him run up the hunter's spear, which has passed through his vitals, until he buries his tusk in the body of the horse, or, it may be, in the leg of the rider."… "The hunter loses his seat at the peril if his life."

Praise of Pigsticking: Pigsticking is the grandest sport that India or any country affords. "Some have condemned pigsticking as cruel, yet of all sports this is the only one practised in modern times where the hunter shares, on almost equal terms, the danger with the hunted. It has a code of honour; the boar is hunted with respect and pursued on certain fixed principles; and there is a *casus belli*, for he is an incorrigible plunderer."

An enthusiast has composed the following imperishable verse:

> *Youth's daring spirit, manhood's fire,*
> *Firm seat and eagle eye,*
> *Do they require who dare aspire*
> *To see the wild boar die.*

Under the altered conditions in India pigsticking is now almost a dream of the past, and all the above of little more than academic interest.

Falconry

Hawking is not now so much practised in Rajputana and Northern India as it was even sixty to eighty years ago. In 1908 an expert modern falconer wrote, "A few days' roaming about a river bank with a net, a set of nooses and some mynahs and sparrows in a cage, and I had collected two peregrines—one a laggar and the other a splendid dark bird in her first year—a *saker*, a *luggar* and two merlins, and within a month was ready for houbara, herons, paddy-birds, crows, kites, hoopoes and larks, and surely it would be a bad day on which I could not find one or other of the above. The *saker* I kept exclusively for kites, the young peregrine was all there when she saw a heron, and both had been 'entered' to houbara."

The list of animals and birds which can be captured through falconry in India is a long one: antelope, gazelle, hares, cranes, egrets, herons, ibises, spoonbills, stone plovers, storks, houbara, florican, junglefowl, partridges, peafowl, sandgrouse, crows, kites, grass owls, vultures, hoopoes, larks, rollers, sparrows.

In the *Journal of the Bombay Natural History Society* "The Review of the Accipitres" is of the greatest interest; and the article, 'Shakespeare on the Noble Art of Hawking' is of interest not only to lovers of Shakespeare.

Fishing

Bombay: The Bombay Presidency Angling Association was started in Bombay with its centre of activities at Powai Lake about 1932 by the late Mr. H.B. Hayes of the American Express Co. Inc., Mr. J.G. Ridland of the Imperial Bank of India and few others, the fishing rights being secured from the Bombay Municipality.

In the initial stages the only fish available were minnows (*Rasbora daniconius*) and olive carp (*Barbus sarana*), but several thousand fry of rohu and catla were released in the lake, and they have grown both in size and numbers. The club is now well established with a membership of over a hundred, and the lake teems with rohu and catla, the former scaling between 20 and 30 lb. while a 65 lb. catla was landed a few years ago. As usual, much heavier fish are said to have got away! More fry have again been recently released and the lake is well stocked for many years.

Madras: The Angler's Club initiated in Madras in 1946 was short-lived. It would seem that India is not yet ready for an Angling Association on Western lines. Perhaps the Angling Club now (1952) proposed to be formed in Mysore State may have more lasting success.

Books on angling in India are listed in my "Bibliography of Big Game Hunting and Shooting in India and the East." (*JBNHS* 49; 222-241).

Trout of the East and the West

The so-called "Indian Trout" of the rivers of Central India, Northern India, Assam and Burma is a worthy rival of the trout and grayling of Western countries. But the successful introduction of the trout of those lands in the upland streams and lakes of Ceylon, the Nilgiri and Travancore Hills, Kulu and Kashmir has brought great enjoyment to many anglers; and there is the landslide-formed Guhna Lake in Garhwal which has proved to be a natural spawning ground and is said to be one of the best trout fishing areas in this country. It is now only two marches from Chamoli where is the terminus of the bus route from Hardwar to Badrinath. Six marches from Tehri is the Dodhi Tal (lake) in Tehri Garhwal where the trout are large and five pounders common.

Regarding the introduction of trout into India pp. 601-3 of the article "History of transplantation and introduction of Fishes in India", by S. Jones and K.K. Sarojini, published in Vol. 50 No. 3 (April 1952) may be seen.

In these days of the motor vehicle the angler in India has quicker access to localities than formerly; and through hydro-electric projects a number of lakes have been formed. On the other hand, these same projects, and certain canal weirs also, have adversely affected migration of important species to spawning grounds, thereby greatly altering some of the rivers and streams of the country to the detriment of the angler and the food supply of the people alike.

From articles in the *Journal*, and earlier angling books and records, it seems that on the whole the angler is not able at this time of writing to have equal success with mahseer in running waters as in former days. Assam has always been a grand province for the angler, but those formerly prolific waters will have been much altered by the recent earthquakes.

The issue in the *Journal* in serial form of the book by A. St. J. Macdonald, "Circumventing the Mahseer and Other Sporting Fish in India and Burma", and its publication by the Society as a book in 1948 was a notable event.

The Society's *Journal* contains close on 300 articles and Miscellaneous Notes on all aspects of fish and fishing both from the angle of sport and of commerce.

In 1907, following the publication in the *Journal* of certain papers (Vol. 17; 637-644) the Society moved the Government of Bombay in respect to legislation for protection of fisheries in Western India; and on 16th January 1908 (Vol. 18; 668-669) addressed the Government of Bombay asking that the expediency of creating a Fisheries Department be favourably considered. That led to useful results in many directions; but from a New Delhi press report of 10th May 1952 it is apparent that even after all these years a great deal remains to be done.

Of the 1,00,00,000 maunds of fish taken from the sea in each year only 32 per cent is consumed as fresh fish. This, says the report, is due to unsatisfactory transport facilities, inadequate supply and distribution of ice and marketing facilities. Other defects are scattered fishing centres and primitive methods of catching, preserving, transporting and marketing. It is emphasized that with proper arrangements and scientific control the fishing industry can make a substantial addition to the country's food resources. So much as regards sea and maritime fishing.

Inland, the activities of the Fisheries Departments have been in recent years principally directed towards stocking of lakes and tanks. Running waters have not received adequate attention. Moreover, the wanton destruction of the nation's fishery resources through use of explosives, fish poisons, capture and waste of fish fry and spawners—has not at all abated, and is getting worse. India should emulate the example of the Philippines where a favourable public opinion in these matters has been brought about.

Defects in regard to running waters have been pointed out by several contributors—Hamid Khan (Vol. 43; 416-426 and Vol. 46; 193-194); Setna and Kulkarni (Vol. 46; 126-132); and there is a valuable article in two parts by Jones (also in Vol. 46) with which is a long reference list. Fishing contrivances in the Hyderabad State are dealt with by Mahmood and Rahimullah (Vol. 46; 649-654); and there is a note by H. de B. Codrington pointing out how much has yet to be discovered in regard to the Mahseer, the premier sporting fish of India.

The illustrated article, also published in pamphlet form, by Spence and Prater on the "Game Fishes of Bombay and the Deccan" is valuable to anglers. Indeed, the Society through its *Journal* has done much to aid and inform regarding the land and sea fish and fisheries of the subcontinent.

Big Game Photography in India

Wild Life Photography is a modern sport of a high order, perhaps more especially in the wide, open spaces of the hills and plains. The man with a rifle has his difficulties; but the sportsman-photographer who has to take his "shot" at a much closer range and bear in mind half a hundred things of importance before he can press the camera trigger has to be a stalker almost in a class apart.

The first book on big game photograpy in India— WITH A CAMERA IN TIGER LAND — was published by Champion in 1927. This pioneer work attracted much attention and was followed three years later by THE JUNGLE IN SUNLIGHT AND SHADOW, by the same author. Many of the photographs in these books were taken at night by automatic flashlight apparatus; so also most of the 120 photographs in the two sumptuous volumes

by Bengt Berg. The article on 'Measurement and Photography of Big Game' by Stockley is good guidance to the shikari-photographer and the sportsman.

Success with his camera in the forests of Burma is illustrated in the two articles by Peacock; and the late Theodore Hubback enriched the *Journal* with five photographs and thirteen pages of valuable information as to the habits and hunting of the Malayan Gaur, or Seladang. This was followed by his article 'Wild Life Photography in the Malayan Jungles' portraying elephant, seladang and sambar at salt-licks together with eleven pages of great value and interest—Apparatus, Hides, Taking the Photographs, Outfit, The Menace of Damp, Animal Psychology. In another article the vanishing Two-horned Asiatic Rhinoceros (*Dicerorhinus sumatrensis*) is fully written up and pictured by Hubback, with additional notes by Prater, the whole forming a monograph on the species.

The finest-ever photograph of a wild bull elephant taken in the Thayetmyo Yomah of Burma by W.S. Thom at a few yards' distance with a 17½ inch Ross Telecentric Lens, and the thirteen pages of this most interesting contribution are a delight to shikaris.

At the present time the Society has valued contributions from the camera and pen of E.P. Gee of Assam.

Wild Life Preservation

About the middle of the eighteenth century the animals of the open country were still in much the same numbers as they had always been, but following the advent of sporting firearms in increasing number, and the pressure on the land through a growing population, the stock of antelope and gazelle all over the country has been greatly reduced—almost to vanishing point in some places. In areas of Upper India where antelope of both species still have a measure of protection on religious grounds they are still in fair number; but outside those special localities they are becoming more and more scarce.

Through the length and breadth of India it is no longer possible for the traveller to view these lovely creatures from railway or motor car window. In that respect there is a lifeless landscape; nor does the former common sight of a stately bustard now delight the eye.

Everywhere the Great One-horned Rhinoceros is now protected. The wonder is that it has survived its relentless pursuit by poachers and the indiscriminate shooting of it by sportsmen in earlier days. The Wild Buffalo should be under strict protection. The tracking and shooting of a solitary bull has afforded genuine sport.

The Indian Gaur, or 'Bison' seems at present to be holding its own; but too many are shot, some are being poached for meat, and the species is subject to cattle diseases—so there is no room for complacency.

It is much to be feared that the Brown Bear of Kashmir and adjacent hill territories is approaching extermination because of its handsome pelt. A male has been measured to be 7½ ft, with girth of 58½ inches. Weight about 500 lb. The Himalayan Black Bear was formerly in great number in Kashmir and Poonch, but much toll of the species has been taken by sportsmen, and also in organized drives on the ground that the animals are not only destructive to crops but often maul and kill the villagers. Owing to its widely extended forest and mountain habitat, the species is not yet in danger of extermination.

Notwithstanding all the poaching and malpractices, there, is still, in some places and due to local circumstances, a fair but decreasing number of sambar, chital, swamp deer, hog deer, and barking deer. Because of its commercially valuable scented pod the Musk Deer is being everywhere slain.

Survival of the elephant where it exists in a wild state is due to the Elephant Preservation Act, 1873, since when proscribed males only may be killed. Recently, in

the Madras State, interested parties have obtained the retrograde step of an alteration in the law to permit of females also being proscribed in the cause of cultivation.

In earlier days wolves were a real menace to the people in many parts of India. Williamson (1780-1806) relates how the troops used to assist in smoking them out of dens, and shooting, trapping and killing them by various methods. The wolf, being a creature of the open country, has been greatly reduced through extension of cultivation; yet, in a few areas, the animal still gives sporadic trouble.

Kashmir in 1924: Of preservation of game in Kashmir, Ward rightly remarked: 'When we consider the difficulties experienced in preserving game in Great Britain we can imagine what has to be faced in the case of tens of thousands of square miles of rugged and mountainous country. It is useless to imagine that poaching in Kashmir can be stopped.' Since then the situation has greatly worsened. Ward's series on Kashmir and the Adjacent Hill Provinces is a complete *vade mecum* for the sportsman-naturalist.

Kashmir and India in recent years: Of the Kashmir Stag it was reported in February 1951 that since the 1947 troubles began there has been rapid disappearance of the species from localities where it was formerly abundant. The 1950 Pakistan report to the I.U.P.N. was that fauna is being rapidly diminished, and that military occupation of certain Himalayan regions has not bettered the situation; while the report from India said that the situation is gloomy and poaching extensive. The India report referred to the whole country and not to Kashmir in particular.

The Role of the Society: The influence of the Society towards Preservation of Game commenced in 1888, and has continued all through the subsequent years, as can be known through perusal of the many references published at pp. 620-22 of Vol. 47, and other contributions in later volumes.

The Society's Honorary Secretaries—Phipson, Millard, Spence, Sálim Ali, Humayun Abdulali and others—and the Curator S.H. Prater have been ever mindful of the influence which can be exercised, and the Editors have given valuable aid through means of a number of important editorials. Had the experienced and expert advice contained in all the above—and in the special illustrated series in five parts compiled by Prater (Vols. 36-8)—been heeded by the various governments, administrations and departments the rapidly deteriorating state of affairs at present existing would not perhaps have come about. But there are many factors and facets in this matter.

In his address to the Society on the 17th March 1930 the President (H.E. Sir Frederic Sykes, Governor of Bombay), remarked that in this country we are confronted with the almost insurmountable difficulty of persuading the masses to have any regard for the principles of wild life preservation; but there is now much more to it than that. Example is more than ever necessary; for a very great difficulty at the present time in India is the increasing number of officials with no interest in sport or natural history, and the rapidly lowering shikar ethics. Even among those who should know better, proper sporting considerations are subordinated to the hunger for meat and the "something-for-nothing" attitude of mind of the man with the gun.

Want of Public Opinion: At the All-India Wild Life Conference held at Delhi in January 1935 it was declared that Indian Wild Life could only be saved by Public Opinion, and that legislation, however efficient, could do little in matters like these without the whole-hearted support of the Public. There is as yet no sign of a proper public opinion while there has been apathy, and even discouragement on the part of the authorities. 'Forests, while saving us from the ravages of flood and famine, can themselves become a menace to cultivation'; and there have been other utterances which are almost direct incitements to users of guns to turn them against wild life. As the present writer has said in letters to the newspapers, 'deer and other wild creatures are just lumps of meat and catchers of votes.'

Laws are enacted, rules are made and forgotten, for there is no continuity of official enforcement and no public opinion to keep them in mind.

India's Vanishing Asset: A comprehensive pamphlet stressing the urgent need for immediate steps towards conservation was printed in January 1948 and widely circulated, with covering letters from the Society and the author (R.W. Burton) to the Governors-General of India and Pakistan, to Prime Ministers and many other high officials; and a précis was circulated through the newspapers and press services all over the country. The pamphlet was printed in the Society's *Journal* (Vol. 47; 602-22) together with a list of 56 references. The Society's notice about it is at p. 792, Vol. 47. 500 copies of a Special Appeal relating to Reserved Forests was also distributed among divisional and other forest officers throughout the country. Later, a Supplement to the pamphlet by the same author (Vol. 48; 290-299) was cyclostyled and similarly circulated.

At no time did it seem that the above impassioned appeals had attracted any attention except for the one Miscellaneous Note [Vol. 48; 388 (1949)] by M.D. Chaturvedi. But there is reason to suppose that sundry measures such as The Bombay National Parks Act, 1950; The Bombay Wild Birds and Wild Animals Protection Act, 1951; the Committee assembled at Delhi on 23rd and 24th July 1951; and now the Central Board for Wild Life appointed by the Government of India to preserve the Fauna of India (Press Note: New Delhi, 11th April 1951) have stemmed from the original pamphlet and other writings. The Hailey National Park and the United Provinces National Parks Act, 1935, resulted from the activities previous to the 1935 Delhi Conference.

A Central Board for Wild Life: This Board was constituted at Delhi on the 4th April 1952 by a Ministry of Food and Agriculture Resolution. It will function through States Wild Life Committees and will meet at least once in two years.

If this Central Board and the States' Committees have before them in correctly summarized form the principal contents of all the main wild life contributions to the *Journal*: the 16th October 1950 thirteen page Memorandum by the writer: and the Address delivered by M.S. Randhawa to the Section of Botany, 35th Indian Science Congress, Allahabad, 1949 ('Nature Conservation, National Parks and Bio-aesthetic Planning in India'), and study and apply all that is practicable in them, there should be good results: but the States Wild Life Committees need to be formed quickly and all that is decided speedily put in motion or results will be of little avail, also too little and too late as has proved to be the case with previous Conferences and Committees.

A Department for Wild Life: It has to be conceded that no such Department will be formed in India—not yet awhile at any rate—but it was counselled by the Society [Vol. 38; 223 (1934)] that there is need for creating a definite agency within the forest department for administering the laws relative to the protection of wild animals. This was supported in the above quoted note by Shri M.D. Chaturvedi, the present Inspector-General of Forests and a Vice Chairman of the new Central Board. A weighty consideration is that the success or failure of game preservation depends upon a wholly trustworthy and impeccable subordinate staff.

National and States Forest Policies: The recently announced Forest Policy for India should have excellent long-range effect on wild life in general; and the C.P. (1st May 1952) Plan announcing 46 recommendations (including game reserves) for management and future development of the Madhya Pradesh protected forests, tree forests, minor forests, pasture lands, recreation forests, fuel and fodder reserves should be a valuable guide to other States and Unions.

South India and the Nilgiris: At the Meeting assembled at Ootacamund on the 7th June 1933 by the Governor of Madras it was decided to form an Association for the Preservation of Wild Life in South India. The project was launched, but within a year proved completely abortive and was never heard of again.

The only bright spot has been the mostly effective preservation of game in the Nilgiris District [Vol. 41: 384-96 (1939)].

Ceylon: In December 1949 Ceylon attained the long sought for Wild Life Department, and the growth of it during 1950 gave hope that at last the menace of the professional poacher and the commercialization of wild life would be halted.

Uttar Pradesh and Assam: The sub-montane tracts of the former United Provinces have always been well stocked with game animals and birds. With some exceptions this obtains at the present time. Let us hope that no Caliph will arise to alter all this.

In Assam there is now a strong movement associated with the names of P. D. Stracey and E.G. Gee. A thousand pities it began too late to save the Manipur race of the thamin from extermination, for there was sufficient warning of what was happening.

Burma and Malaya: In spite of vigorous efforts and warnings by Smith, Peacock, Weatherby, and Hubback important species have vanished or are nearing extermination in these countries—and this before the two countries were overrun by the Japanese during the last war.

Education in Schools: 'The youth of today must become the conservationists of tomorrow.' The Bombay Natural History Society has worked towards this end with, as yet, no widely extended results, and the present writer has been urging the need for the past five years. Sir Frederic Sykes (1930) said that we should aim at teaching the children to appreciate the value of wild life. In his address to the Ceylon Game and Fauna Protection Society on 14th December 1950 the Governor-General and Patron of that Society said, among other things, 'There is need for extensive propaganda and education, and the Government and this Society can co-operate to convince the young generation in the schools that they will, and must be, the future custodians of wild life'—to which can be added 'and of the forests also.'

At the present time the International Union for the Protection of Nature is making considerable effort in this direction, and Italy, Greece, French Cameroon, Mexico, Belgium, Belgian Congo, Madagascar, and Turkey are issuing special lessons on the subject for the interest of educators and used by teachers and pupils in primary and secondary schools. "In spite of its importance to mankind, the theme of these lessons is little known or totally ignored by contemporary nations." How very true it is that, "Many are the paths along which man proceeds to (his own) destruction....."

■ ■ ■

The Golden Tree-Snake

(*Chrysopelea ornata*)

By Major Frank Wall, I.M.S, C.M.Z.S. (Indian Army)
Year of Publication: 1908

To **Colonel Frank Wall** we are indebted more than to any other man for our knowledge of the habits of the Indian snakes. As a member of the Indian Medical Service he arrived in the country in 1894, where he spent most of the next 30 years of his life and in the course of his duties was stationed in most parts of the Peninsula including Sri Lanka and Burma. Wherever Wall went he collected and studied his material, and by his enthusiasm induced others to collect for him. He was not a museum worker. His interest was in the living creatures and his voluminous writings deal almost entirely with their habits and structure. His larger works include 'The Snakes of Ceylon', 'The Poisonous Snakes of our British Indian Dominions' and 'A Popular Treatise on the Common Indian Snakes.'

He saw active service in the First World War and was with the expeditionary force to Mesopotamia. He also served in France. He was twice mentioned in dispatches and in 1915 received the C.M.G. for his work at Bologne.

Nomenclature. — Scientific.- The generic name is from the Greek *chrysos*, gold, and *peleios*, black. The specific is from the Latin *ornatus*, adorned.

English. — The golden tree-snake or the gold and black tree-snake.

Vernacular. — "Kalla Jin" the name given by Russell for a specimen probably obtained in Bengal, is probably Urdu implying "black saddles" with reference to the black cross bars. Its name in Ceylon according to Ferguson is "pol mal karawala."

Dimensions. — The largest specimen I know of is the one obtained by Evans and me in Rangoon, which taped 4 feet 5½ inches. Specimens over 4 feet are unusual.

Physiognomy and bodily configuration. — The snout is broad, blunt, and rounded, the head flattened and the neck moderately constricted. A moderately well developed eye with golden iris (Cantor says black) gives a vivacious expression to a not unpleasing countenance. The pupil is round. The nostril is moderate in size, and placed entirely or almot entirely in the anterior nasal shield. The body though rather slender is far less so than in other tree snakes, notably *Dendrophis, Dendrelaphis* and *Dryophis*. It is rather depressed. The belly is peculiar in the ridges on the lateral aspect of the ventral shields. The tail is long, and tapers very gradually. It is about one-fourth the total body length and it is ridged beneath on either side similarly to the belly.

Colour. — Eight colour varieties, many of which I have not seen.

Identification. — This is an easy matter if attention be directed to scale characters.

The ridged (keeled) condition of the central shields taken with a vertebral row of scales in no way different from its adjacent rows, establishes the diagnosis. The ridged character of the ventral shields demands some qualifying remarks. It must be impressed upon the student that the keels in *Chrysopelea* are sharp and pronounced, with a minute notch on the free edge of the ventral corresponding to the keel.

This character of keel is only seen in two other genera, viz., *Dendrophis* and *Dendrelephis*, both of which are tree snakes also. In both these genera however the vertebral row of scales is enlarged, and hexagonal as in the kraits (*Bungarus*). It is to be noted that several other snakes have what may be called false keels on their ventrals, that is, the ventrals are laterally angulated. False keels are rather rounded (obtuse), and have no notch at the free

edge of the ventral. The outlines in section may be compared roughly to those of a punt, and a dinghy.

It is interesting to note that these false keels are to be seen chiefly in those snakes which manifest a climbing habit. For instance, in the genus *Lycodon*, witness the species *aulicus*. In the genus *Zamenis*, observe the ventral in *mucosus*. In the genus *Coluber* in notably the arboreal species *prasinus*, *frenatus*, and *oxycephalus*.

Haunts. — Very opposite opinions prevail as to its haunts. Cantor says it is seldom seen in trees, but more frequently on the ground in the grass. Stoliczka supports this observation, and says though he caught several specimens in the grass, or between low bushes, he but once saw one actually on a bush. Flower on the other hand says his experiences are very different in this as in other matters from those related by Cantor, and he agrees with Günther's suggestion that the reason it is not more often seen in trees is because it makes a too rapid retreat.

I am able to support both parties, for I have seen it high in a tree on a naked limb, and on several occasions on bushes, or on the trellis work about tennis courts and verandahs. I never met it on the ground myself, but many of the specimens brought me were reported on the ground. I have not the least doubt that the species is essentially arboreal in habit, but this does not prevent it making frequent excursions to the ground either in search of a fresh feeding area, or in the pursuit of the incautious quarry which its keen vision has detected from its exalted station amid the branches overhead.

It is only natural that it should be more frequently encountered on the ground, because the eyes of the pedestrian are directed below the level of his head, even at his feet. Men other than bird-watchers, fern and orchid hunters, and such like do not gaze much aloft, and the snake reclining along a branch or on the top of a trelliswork even about one's own height, will frequently escape detection though but a few feet or even inches away.

ASHOK CAPTAIN

The very fact that Cantor and Stoliczka in unison with other observers mention that geckoes are the principal food seems to me to refute their suggestion that *Chrysopelea* is terrestrial in habit, for geckoes are eminently arboreal. It is to be noted also that all the food partaken of, other than geckoes, is of a nature to be obtained by climbing only. Cantor's inclusion of frogs in their dietary does not vitiate this remark, for though he does not say so, the species taken may have been arboreal forms only.

Chrysopelea is not infrequently found about, and actually inside habitations. Flower mentions this, and Evans and I had similar experiences in Burma. I well remember in Colombo, too, one that had taken up its quarters in an old packing case which was full of straw and other packing material. A cooly was ordered to clear this out, and stepped into the box to carry out his orders. His exit reminded one of an incautious bather who has stepped into over-hot water. The alacrity of movement so foreign to the cooly's nature was explained by the subsequent discovery of a snake of this species.

Disposition.—Very divergent views again have been expressed on this point. Cantor remarks on the gentleness of the species, whilst Flower on the other hand says "*Chrysopelea ornata* is the fiercest snake I have met. Under circumstances when most snakes, harmless and poisonous alike, would try to glide away quietly, this one will turn to attack the person who disturbs it, and will attempt to resist capture to the uttermost, striking, and biting ferociously. ...Individuals I have at various times tried to keep in captivity showed no signs of becoming tamer, and would always bite my hand when I put them in the vivarium, and being also an annoyance to the other inmates of the cage, I have only kept them for a few days at a time."

I must say I can abundantly confirm Flower's experiences. There is no doubt that this snake is decidedly plucky, and on occasions fierce, but I would not suggest that all are equally vicious. I think that snakes, like other creatures, exhibit individual character.

I well remember my servant in Rangoon trying to effect the capture of a large specimen in a hedge adjoining my compound. I arrived on the scene when the excitement was at its height and discovered that all the menial establishment amounting to ten or more had been pressed into service. The snake had fought most courageously to elude capture, and struck at any one who ventured to attempt to grasp it. My boy, confident of master's solatium in the form of a rupee if the creature was captured alive, had been struck at and bitten, and I hardly knew which to admire most, the servant's determination and courage or the snake's vigorous endeavours to retain its liberty.

Flower mentions one in a fit of rage biting itself with such vigour that its teeth were fixed into the side of its body, and I can remember recapturing one which had escaped from my vivarium and had taken refuge between some boards in my house. When extricated after some difficulty, and with the employment of some force, it struck at and buried its teeth in its own body.

The fact that this snake will even face, much less try conclusions with a tuctoo (*Gecko verticillatus*) is eloquent proof of its intrepidity. Flower says: "I have known it eat *Hemidactylus frenatus* and *Gecko verticillatus*; the latter may give battle to the snake for some hours before being finally swallowed." I have two instances of the indomitable courage of this giant gecko. In one instance it was clearly the aggressor, and not only confronted but actually drove back a large rat snake (*Zamenis mucosus*), a species whose courage is well established, and actually during the retreat snapped and bit off part of its tail. In the other instance witnessed and recounted to me by Captain Lloyd, I.M.S., on Sandoway Island when this gecko was in conflict on the ground with a *Chrysopelea ornata*, it would be difficult to say which was the aggressor, but it is probable that the snake engaged the lizard, not expecting to meet a foe of such prowess.

Food.—*Chrysopelea*, whilst showing a decided partiality towards lizards of the family Geckonidae, accepts with avidity many other creatures that chance has to offer.

Members of the genus *Hemidactylus* (Geckos) are most frequently found to have furnished the meal, obviously from the relative abundance of the species in this genus and the numerical strength in individuals of many of the species which frequent trees. Many other lizards, however, fall victims to its voracity. Cantor mentions frogs as well, but I have never known one taken, have known them refused in captivity where lizards were accepted, and Mr. Millard tells me also he found frogs were not acceptable. Among other lizards Cantor found *Ptychozoon homalocephalum* (Gliding Gecko) taken once, and Evans one of the flying lizards (*Draco toniopterus*). I have known *Calotes versicolor* (Garden lizard) taken, and Flower the giant gecko or tuctoo of Burma (*Gecko verticillatus*). Evans and I noted that one had eaten a bat (*Taphozous longimanus*), and Evans has since recorded

ASHOK CAPTAIN

two instances where bats were devoured. In captivity it "feeds freely on bloodsuckers, sparrows, geckoes, and mice, but never eats frogs," killing by crushing in its folds.

It is interesting to note that a pet *Chrysopelea ornata* was fed with milk out of a saucer, the snake being held near the head and the saucer put to it, when it readily drank the milk, and in comparatively large quantities.

Habits. — The striking beauty of this snake, whether seen reclining or moving in its native haunt, could hardly fail to arouse the keenest admiration in the breast of the most unappreciative and phlegmatic disposition.

I watched with admiration recently the adroit, though stealthy, manner in which one in captivity in the Colombo Museum balanced itself, and moved along my walking stick though this was more slender than the snake itself.

This snake has the habit of clinging to the trunk of a tree, head downwards, in a very extraordinary manner, and I have seen it under almost exactly similar circumstances. My specimen was stationary, clinging (one could not say reclining) head downwards, about 30 feet from the ground, on a large bare trunk, which rose almost perpendicularly. I marvelled at the tenacity of its grip in such a situation. It had thrown its body into a very wide S across the limb, and it strikes me now very forcibly, after reading Flower's and Shelford's observations, that it may have been "gathering itself" for a leap. The enraptured observer will be even more captivated with the grace and agility attending its movements from branch to branch, and the consummate ease with which it will scale a perpendicular trunk. Its flash-like disappearance aloft without apparent effort must be witnessed to be fully appreciated. I very much doubt whether any snake moving along the flat displays greater speed than this species in its arboreal environment.

But its marvellous attainments do not end here, for this snake is endowed with the capability to spring, or "fly" as some prefer to call this jactatory effort. Here one is forcibly reminded of the eulogistic terms in which the late Professor Owen summed up the athletic performances of these limbless creatures.

He says: "They can outclimb the monkey, outswim the fish, outleap the jerboa, and suddenly loosing the coils of their crouching spiral, they can spring into the air and seize the bird upon the wing."

One has only to be acquainted with *Chrysopelea* to realize that Owen's words convey no fulsome flattery. That it actually can spring is vouched for by more than one reliable observer. Flower in 1899 reported having seen "a small one, about 2½ feet long, take a flying leap, from an upstairs window, downward and outward on to a branch of a tree and then crawl away among the foliage. The distance it had jumped was measured, and found to be nearly 8 feet."

Curiously enough in the very month (May) and year (1899) when this record of Flower's was published, Mahon Daly wrote from Siam reporting his having witnessed a similar feat and though he could not identify the snake he said that he and his Karen interpreter saw a snake, "about 2½ feet long, sail from a very high tree on one side of the road to a lower one the opposite side."

In confirmation of these very extraordinary aerobatic feats which I have no doubt many might be inclined to disbelieve is the report made by Shelford of similar performances. This observer relates that three native witnesses in Sarawak made a similar statement on three different occasions independently of one another, and at considerable intervals of time.

This was to the effect in each case that the snake had been seen to "fly" from some height to the ground beneath. In all cases the snake was reported to have kept its body rigid during this feat, and to have met the ground at an oblique angle. In one case the snake proved to be *Chrysopelea ornata*, in the second instance a snake of the same genus, viz., *C. chrysochlora*, and in the third *Dendrophis pictus*.

Shelford calls attention to the fact that all these snakes are alike in the peculiar ridged condition of their belly shields, and he made experiments to ascertain the truth of these reports. He says: "A specimen of *Chrysopelea ornata* was taken to a height of fifteen to twenty feet, and allowed to fall several times; after one or two false starts the snake was felt to glide from the experimenter's hands, straightening itself out, and hollowing in the ventral surface as it moved, and it fell not in a direct line to the ground, but at an angle, the body being kept rigid the whole time. If the snake was thrown up into the air, it seemed unable to straighten itself out; it had to be launched, so to speak, from the hands in order to induce it to assume the rigid position."

He implies therefore that these "flights" are not accidental falls but deliberate voluntary efforts, and suggests that the hollowing of the belly between the two ventral ridges may act mechanically after the manner of a parachute, impeding the action of gravity, and buoying up the creature so as to reduce the momentum with which it would strike the ground. He illustrates this point by comparing the fall of a piece of bamboo bisected longitudinally, and the concave face downwards, with that of a piece of bamboo in its cylindrical form. In the former case the descent is retarded. I prefer the use of the term "springing" to that of "flying" in describing these feats. Its only rivals in acrobatic and scansorial achievements are the tree snakes of the genera *Dendrophis* and *Dryophis*.

Breeding. — Our information on this point is scanty. Evans and I obtained one in May with ovarian follicles impregnated, one in June with 9 eggs in the abdomen, and a small specimen, length not noted, believed to be a hatchling in June. These were all obtained in Rangoon. Evans acquired a specimen from Hanthawaddy, Lower Burma, in June containing 11 eggs; and a brood of 6 young in June in Rangoon. It is clear from the above notes that it is not a very prolific species. The measurements of the eggs were not recorded. The young in the brood recorded by Evans measured from 4½ to 6 inches in length.

It is not known whether this snake is oviparous or viviparous. Without being too positive I am inclined to think that the eggs I extracted "ex abdomina" contained embryos in a very early stage of development.

This snake grows 9 or more inches each year, so that the specimens reported by Evans 13½ and 14 inches long in August were the previous year's production.

My smallest prospective mother was 3 feet 7 inches long in June and therefore in her 5th year.

Distribution.—This is very extensive, ranging as it does between the western shores of India on the extreme West, through the Malayan regions (continental and insular), South China to the Philippines in the extreme East. So far as the Indian Peninsula is concerned its distribution is peculiar, and very interesting. It is only found in a small tract of country in the southern part of the Malabar Coast, and in Eastern Bengal.

Golden Tree or Gliding Snake *Chrysopelea ornata* (Shaw)
1,5. Head dorsal view; 3. Head side view; 4. Ventral view; 2,6. Patterns of back

In Sri Lanka it is not very uncommon, I met with more than one specimen in a four years' residence though not at that time a collector of snakes. It is referred to by Ferguson, Haly and Willey from the plains. Mr. E.E. Green tells me he has never heard of it in the hills in that Island, i.e. above about 1,500 feet.

It is a fairly abundant species throughout the Malayan Region, and extends throughout Burma. In many parts of Lower Burma it is a common snake (Rangoon, Pegu, etc.) In the extreme south of this Province it has been recorded from Mergui and Tavoy Island (Sclater). Captain Lloyd, I.M.S., captured a specimen on Sandoway Island which I have already referred to . To the east of this Province Sclater has recorded it from Moulmein, and Evans and I had specimens sent to us by Colonel Bingham from the Southern Shan States. The British Museum has a specimen from the same donor from the Ruby Mines, but it appears to be uncommon in that part, for at Mogok Mr. Hampton tells me he has had no specimen in a 9-years' residence.

From Burma it extends to the North, through Assam, and across the Brahmaputra into the Eastern Himalayas, and in a westerly direction into Eastern Bengal, where its exact limits are somewhat uncertain.

It occurs within the Gangetic Delta (Calcutta and Barisal) and it is probable that its western boundary is defined by the Hoogly and Teesta rivers. It has not as yet been recorded from the Andamans or Nicobars, but Annandale refers to a specimen taken on Narcondam Island by Major Anderson.

■ ■ ■

Butterflies of India

Brigadier William Harry Evans, D.S.O., R.E., F.Z.S., F.E.S. **(Indian Army)**
Year of publication: 1922

I do not think that anyone, who has taken the trouble to look, can fail to be impressed by the splendour of the butterflies of India. There are many who would like to study them further, but very few proceed beyond mere admiration, chiefly because of the difficulties involved in following a pursuit without a guide. It is the purpose of this article to draw attention to the variety of interests that underlie the hobby of butterfly collecting and to explain how these beautiful insects may be captured and preserved. I think it was A.R. Wallace who wrote somewhere that the story of evolution is written on the wings of butterflies and I believe that, when the study of zoology can be correlated with the studies of geology, geography and botany by a superman of the caliber of Darwin, the mystery of evolution will be solved. Before the solution can possibly be reached, and it will not be in our generation, a mass of observation work has to be done. The professional zoologist ignores the butterflies and devotes himself either to the lesser known groups or to such as have a definite economic value; he admittedly leaves butterflies to the amateur and so here is a field in which the amateur, who happens to be an observant student of nature, can help on the attainment of knowledge.

The Indian subcontinent, wherein for zoological purpose are included Ceylon and Burma, is probably the most ideal country in the world for pursuing the study of butterflies. It offers the extremes of heat and cold, of dampness and dryness, of desert and rank jungle, islands and continental areas, an ever-varying vegetation and in many parts sharply marked seasons, while its geological history is most interesting. For faunistic purposes the world is divided into the American. African, Palaearctic and Oriental regions. With the American region we are not concerned. With the African region we have a connecting link through Baluchistan, but in former years the connection must have been a much more important one, since there are a number of genera which we share with Africa. The Palaearctic region comprises Europe, the Mediterranean littoral of Africa, Western and Northern Asia. As far as we are concerned this region is divided into the European sub-region, which embraces Western Asia and reaches us through Baluchistan; the Central Asian sub-region, reaching Chitral and to a less extent the N.W. Himalayas; the Chinese sub-region, which enters the Indian between Sikkim and Northern Burma. The Oriental region comprises the Australian and the Indo-Malayan sub-regions and the latter is divisable into the South Indian (including Ceylon) and the Malayan, covering all Lower Burma; it must be remembered that the

W.H. Evans, one of the most distinguished entomologists of our time, died in November 1956 at the age of 80. He was born in Shillong, Assam, in 1876, his father General Sir Horace Evans being there in command of the 8th Gurkha Rifles.
His first contribution to the *Journal* of the BNHS was in 1912 and he sent articles on the Identification of Indian Butterflies, including those of Burma and Ceylon, during 1922-26. These were published in book form in 1927, a surprising achievement for one who had always worked hard in his profession as a Royal Engineer.
He was commissioned from Sandhurst in corps of Engineer and awarded the DSO for gallantry during WWI in France. Retired as Chief Engineer, HQ Western Command in 1931.

mighty Himalayas are mere children on the face of the globe and only possess a bastard fauna made up of immigrants from other far older areas.

Butterflies differ from moths in a number of ways, but no hard and fast line can be laid down. The chief distinguishing characters are:

a.   Butterflies for the most part fly by day; they never fly at night, but a few species remain dormant until the dusk.

b.   They have as a rule knobbed or hooked antennae, which are straight and are held in front of the head.

c.   Most butterflies rest with the wings erect.

d.   The upper and lower wing on one side of the body are never joined at the base.

Yellow Swallowtail *Papilio machaon*

The species of butterflies are divided into a number of families, sub-families and genera depending on their structure, into the details of which we need not enter here. The Indian butterflies have not up to the present been given English names, though every school in the hills has assigned to the more conspicuous species found in the immediate neighbourhood fancy names of sorts. Scientifically a butterfly is recognized by two names, that of the genus being followed by that of the species; for instance *Papilio machaon* is the ordinary English swallowtail. Butterflies that occur over a large area are apt to develop into a number of well defined races and in order to distinguish them a tri-nomial system of nomenclature has been adopted; for instance *Papilio machaon* is the European race of the common yellow swallowtail, *Papilio machaon asiatica* is the Western Himalayan race, *Papilio machaon sikkimensis* the Chumbi Valley race and so on. If it is desired to distinguish a variety, the abbreviation v. is put before the varietal name; for instance *Papilio machaon asiatica* v. *ladakensis* is the short-tailed variety of the common yellow swallowtail found in the Western Himalayas. If the variety is confined to only one sex, the sex sign ♂ for male and ♀ for female is put before the v.

For the purpose of this article the main divisions into which the butterflies, or scientifically the Rhopalocera, are divided may be briefly described as follows:

A. **Danaids**. Large, tough, insects which contain nasty juices and can emit evil odours at will, whereby they are rendered distasteful to their enemies. Two tawny and one blue and black species of the genus *Danais* and one white spotted dark velvet brown species of the genus *Euploea* are to be found lazily flitting about every garden in the plains. In N.E. India and Burma many of the *Euploeas* are shot with a most splendid iridescent blue colour. A third genus (*Hestia*) containing very large black-spotted diaphanous butterflies is to be found near the coast in (Ceylon) and South India and in the mangrove swamps of Burma:

Chocolate Tiger *Parantica melaneus* — a Danaid

ISAAC KEHIMKAR

B. **Satyrids or "Browns"**. As a rule the members of this somewhat numerous group prefer the shade and are most often to be seen flitting about in jungle; a few species patronise rocky slopes. They are mostly sober coloured insects with rings or eyes on the wings. They vary in size from the tiny *Ypthima* to the very large *Neorina* of N.E. India.

C. **Morphids**. This group attains its greatest development in South America. They are only found in or on the edge of thick jungle and do not fly much by day unless beaten up. Nearly all the species are very large and, though our Indian forms do not equal the S. American ones in splendour, yet they have nothing to be ashamed of.

D. **Nymphalids**. These are the. true sun lovers amongst butterflies and the group contains a large number of handsome species. The well known "Painted lady" can be taken as their universal representative, but the diversity of forms is extraordinary. Some of them, the genus *Charaxes* for example, have very large and strong bodies and can fly like birds; others, such as the delicate "Map butterfly" sail gracefully in the sunshine. The wonderful "Leaf butterfly", the white and red "Admirals", the "Tortoises", the "Purple emperor", and the "Fritillaries" are all members of this group. We all know the merry little bright blue and yellow *Junonias* that flit about just in front of us along our bungalow paths, also the large *Hypolimnas* with blue-ringed white circles that often appears in swarms shortly after the break in the rains.

E. **Papilionids or Swallowtails**. Many of them have no tails and are mistaken for members of the preceding groups; however a glance at the legs settles the point at once; in this and the following groups the forelegs are as long as the others, while in the preceding groups the forelegs are short and quite useless for walking. There are many magnificent swallowtails in India and they can vie in beauty and diversity with their cousins in any other part of the globe. In the South and North East and in Burma there fly the *Ornithoptera*, great black insects with brilliant yellow hindwings, which fly slowly far out of reach at the tops of trees. Then there is the wonderful black swallowtail with a peacock-green hindwing found in the hills, and the delicate white, black striped, swallowtail of the Himalayas. In the family are included a few species belonging to genera other than the true Papilio. There is the *Armandia*, a truly magnificent butterfly from Bhutan, the Naga and Chin Hills, a many-tailed creature with a large, red area on the hindwing. The *Teinopalpus*, a wonderful green and yellow butterfly, that is to be found on Tiger Hill, Darjiling, and the hills of Assam and North Burma. The *Leptocircus*, a small and very curious looking insect, which has enormously long tails and presents a striking resemblance to a dragonfly. Finally the *Parnassius* or "Apollo" butterflies, inhabitants of the highest Himalayas; beautiful white insects with black, red or blue spots.

F. **Pierids or "Whites"**. The majority are white, such as the well known "Cabbage white", but many are yellow and a few are even red or blue. There are the "Brimstones", "Clouded yellow" and "Orange tips". The most characteristic Indian

Banded Treebrown *Lethe confusa – one of the Browns*

Jungle Glory *Thaumantis diores* –a forest dwelling Morphid

representatives are, perhaps, the members of the genus *Catopsilia,* large greenish white insects, and the small yellow *Terias*, several species of which swarm in every garden. The most variegated species belong to the genus *Delias*, the plains member of which is a white insect with larger red spots along below the edge of the hindwing.

G. **Lycaenids or "Blues".** The most numerous family of all and in many respects the most interesting. The diversity in colour, markings and shape is greater than in any of the preceding groups. The Lycaenids are divisible into two main groups, the "weak" and the "strong" or the true "blues" and the "hairstreaks". Amongst the "hairstreaks" are to be found the most brilliant metallic blues, greens and brassy tints. In many species the tails are very long and in others short and thread-like.

H. **Hesperids or "Skippers".** This group differs greatly from those that have preceded it. They are mostly small, rather dull coloured, insects, with large heads and bodies and a very rapid flight. The group is not so well known as the rest, but the advanced entomologist finds that it presents features of very great interest. The accumulation of a great deal more material is needed to enable the group to be worked out properly.

Below is a table, which shows how the various groups are distributed amongst the sub-areas detailed in para. 2. The totals given in the last column are for the actual species known; many of these species have developed into more or less well defined races according to the areas they occupy and the modern tendency is to name more races as more material becomes available.

With the Nymphalids are included two minor groups, which are closely allied viz., the *Libythacinoe* and the *Nemeobiidoe*.

As no doubt everyone knows, a butterfly has passed through the stages of egg, caterpillar and chrysalis. The study of the early stages of the butterflies of India has with two brilliant exceptions been most woefully neglected. The first of these is Mr. T.R. Bell, the author of the papers that appeared in the *Journal* entitled "Common Butterflies of the Plains"; he has discovered the life history of practically every butterfly

Fluffy Tit *Zeltus amasa* – one of the Blues with long tails

Group	Sri Lanka	South India	Baluchistan Sind Pakistan	Chitral Ladakh	N.W. Himalayas	N.E. India N. Burma	S. Burma	Andamans Nicobars	Total
Danaids	11	11	1	3	7	18	28	18	37
Satyrids	16	28	10	17	51	117	57	8	189
Morphids	1	3	-	-	-	17	16	2	20
Nymphalids	38	49	6	30	88	180	145	33	247
Papilionids	15	19	4	10	25	68	49	11	91
Pierids	30	36	19	18	39	51	36	17	83
Lycaenids	72	86	14	44	80	237	233	61	470
Hesperids	4	78	7	13	58	198	154	33	278
Total	187	310	61	135	348	886	718	183	1,415

Common Awl *Hasora badra* — Skippers have large heads and bodies

that inhabits the North Kanara District of Karnataka; as a matter of interest I may mention that in at least one case Mr. Bell has reared a butterfly that has never yet been found flying. The second is the late Mr. P.W. Mackinnon, who in Vol. XI of the *Journal* described the life history of most of the butterflies of Mussoorie.

Every fully developed living organism is said to climb up its genealogical tree before it attains maturity, so that the importance of studying the early stages can hardly be over estimated; our classification still presents many imperfections, which will not be removed until a great deal more has been discovered regarding the early stages. Breeding butterflies is a rather troublesome business and necessitates continued residence in one place; the opportunities afforded the ordinary official are somewhat limited and it is chiefly to the planter or the retired individual that we must look for assistance. The great desiderata are observers in the Darjiling district and in Assam and Burma more especially the rubber planters or tin miners of Mergui and Tavoy. Many years ago a distinguished American naturalist, who spent some time collecting butterflies in India, evolved a classification for the Skippers based on the eggs, but lack of material rendered it unreliable. Under a microscope of moderate power the eggs are wonderful things. The caterpillars are of many different forms and colours, while the chrysalis are often most curious and are concealed in a marvellous manner. I will not enlarge on this subject, but will refer those interested to Mr. Bell's articles, than which nothing better has ever been published. A study of the food plants in various localities with notes as to how they differ would be most valuable information towards affording an explanation of the geographical variation of butterflies.

The investigation of the structure of the perfect insect can safely be left to the cabinet naturalist, but it is only the field naturalist who can supply information regarding the habits of butterflies. There are numerous points to be noted. Males are often to be found playing about in the sun or sucking moisture in damp spots. The female attends more strictly to her business of egg laying and requires watching as to where she lays her eggs, whether singly or in clutches, on what particular food plant and whereabouts on it. Caterpillars can sometimes be seen feeding openly, but for most of them a very close examination is needed; many are night feeders; a good plan is to beat bushes and with luck the caterpillars will fall into a net held below. The caterpillar passes through various

moults and requires examination at every stage. The act of turning into a chrysalis demands close observation and the manner in which the chrysalis reposes; some hang free from a twig, others are secured by a girdle, while some remain like a grub inside a fruit. The actual emergence of the butterfly is a sight rarely seen but very well worth watching. The habits of the actual butterfly require much observation; the season of emergence, number of broods in the year, duration of life, mode of flight, nature of habitat and so on all require recording. There is no end to what can be done and all of it is interesting. There seems to be a very general impression that the life of a butterfly is limited to one day; if an enemy secures it on the day it emerges, then all is over, but many survive for very long periods; for instance in the Himalayas nearly all the butterflies one sees in the Spring emerged in the Autumn and have lived through the Winter, coming out very often for a flight on a warm day; in Japan there is a species that emerges in July and flies till the following May.

If a butterfly flaps about in a lazy fashion, it is probably more immune from dangers than those that have a rapid flight or seek cover rapidly. This brings us to the interesting subject of mimicry and in India we have some of the most striking examples in the world. Now the Danaids are probably the most highly protected family and we find that the female of a common Satyrid and a Nymphalid resemble a certain common Danaid almost exactly, though the males in each case are totally different. Again among the swallowtails the red-bodied group are highly offensive and we find that certain females of the black-bodied group resemble them exactly except for the colour of the bodies. Again among the *Euploea* genus of the distasteful Danaid family there is a marvellously close resemblance between species inhabiting the same area; in South India, for instance, the only three members of the genus are so alike that somewhat of an expert is required to distinguish between them. The reason for this latter form of so-called mimicry is stated to be that the young enemy has to learn by sad experience what is distasteful and what is palatable; so he starts by sampling everything that comes along and soon learns to distinguish the nasty from the tasty; thus the nasty group are likely to lose fewer individuals if they present the same general appearance to the former. For the same reason it is an advantage for a distasteful butterfly to be coloured conspicuously; it is thus able to warn the enemy that it should be left alone.

The observer will soon notice that a large number of butterflies, more particularly the inhabitants of jungles or undergrowth, are marked in a peculiarly cryptic fashion on the underside and that, when at rest, they are so assimilated to their surroundings as to be practically invisible. Many of them are wonderfully coloured above and cannot fail to attract attention but, when they alight, they seem to disappear. The famous "Leaf" butterfly can be cited as one of the most striking examples; the upperside bears a broad yellow or blue band, while the underside almost exactly resembles a dried leaf.

Orange Oakleaf *Kallima inachus* – a classic example of mimicry in nature

The tails of a butterfly are considered to be an imporant life saving device and certainly they do not seem to be much use for anything else; the hind wings project well beyond the body and just before the tail there is an eye; the idea is that the enemy thinks that the hinder end is the head with the prominent eyes and that the tails are the antennae; he makes his dart, but the butterfly flies cheerfully away minus his tails the loss of which do not worry him at all, but of course he will not escape so easily again.

Several theories have been propounded to explain the mystery of mimicry; to me the whole subject remains an absolute mystery and I can safely assert that a great deal more observation work is needed before anyone can produce a theory that will convince the man in the street. I cannot believe that a butterfly has been able to perfect the art of camouflage, as he undoubtedly has done, by means of his own unaided intelligence.

Bound up with mimicry is the question as to what are the enemies of butterflies and here again much observation is needed. I think that the principal enemies are lizards, as far as the butterfly is concerned, but there are undoubtedly many others. I have seen a kingcrow making a good meal off butterflies on the wing, and a bush containing a praying mantis is often marked by numerous butterfly wings strewn below it. In the earlier stages the enemies are probably much more numerous, and birds in particular eat the caterpillars very freely; many of us as schoolboys have been bitterly disappointed to find an ichneumon fly emerge from a chrysalis we have carefully reared.

Butterflies share with all other living beings the strong natural tendency to enlarge their sphere of action, but certain species are known to indulge in the most extensive migrations. Certain Pierids (the *Catopsilias*) are in India the most persistent emigrants. They can sometimes be seen in great herds trekking in a straight line over hills and plains at a pretty constant speed. Certain other species often join in the migrations. A common blue (*Lampides boeticus*) migrates in the Spring. In March at Rawalpindi I have observed this species migrating for several weeks; across a width of 20 yards I counted 90 per minute passing during the period of greatest intensity; throughout the whole period, the direction was 5 degrees North of West. Again and again I have noticed butterflies flying in a definite direction; they may turn aside to a flower or to inspect a passer-by, but they eventually continue on in the same direction. In the *Journal of the Entomological Society* there have appeared recently some interesting articles dealing with the migrations of certain Pierids in the Island of Trinidad; the butterflies apparently started at the South end of the island, travelled direct to the North end, then turned West. leaving the island eventually in this direction for the mainland of America. The density of these migrations was so great that motor cars had to stop moving on the roads, since nothing could be seen through the wind screens.

I should now like to touch lightly on the subject of variation, regarding which much has been written, but there is no doubt that theory still predominates over proved facts. The whole subject is most fascinating and, as I said at the commencement of this article, not only do butterflies offer the best medium for its study but also we are better situated for the purpose in India than anywhere else in the world. There are four kinds of variation to be thought of: sexual, individual, seasonal and geographical. Sexual variation can be dealt with very briefly. In many species the two sexes are almost indistinguishable; in the majority there is a marked difference; in a few there is no resemblance whatever between the sexes and it is only comparatively recently that they have been ascertained to belong to the same species. Among the nasty groups the resemblance between the sexes is pretty close and this is also the case more or less amongst the group with cryptic undersides. In the unprotected groups there is usually a considerable difference; it is the female that is always duller and less conspicuous than the male, while in certain genera the differences

between the males of the various species are considerable, but the females are so alike as to require a first class expert to allocate them to their correct males. It is no doubt the object of Nature to preserve the life of the female for a longer period than the male, since after fertilisation she has to devote her time to laying her eggs. Possibly Nature's intention is that the bright coloured males in the unprotected groups should be sacrificed to the enemy, so that the insignificant females are preserved.

Individual variation may arise from a number of causes; different nutrition in the caterpillar stage, effect of light or temperature on the chrysalis, are probably the most important, but there are no doubt others that we do not even suspect at present. In some species the variation is extraordinary; for instance the undersides of no two leaf butterflies are alike. Again in the little yellow Pierid (*Terias*) of the Indian garden the variation between individuals is tremendous. In many butterflies the variation is very small, but a minute examination will show that no two individuals are exactly alike any more than are two human beings. All the varieties of this class can be more or less linked together by intergradations, provided one can obtain a long enough series. There is, however, another class of individual variation, where very distinct varieties occur, that cannot be linked together; this kind of variation is called dimorphism. For instance a few species of Papilionids exist in two totally different forms in both sexes. Two other Papilionids each have three totally different forms of females. In one of these, the black-bodied *Papilio polytes*, the first form of female resembles the male and is comparatively rare; the second resembles a rusty red-bodied swallowtail, which only occurs in the peninsular portion of India, where alone this form of female is to be found though the species is met with nearly everywhere; the third and commonest form is like another red bodied swallowtail found everywhere. There is yet another kind of individual variation known as an aberration or sport; these are rare but occur in some species more than in others; they are usually very different to the parent form. This class of variation has attracted a great deal of interest of late years and forms more or less the basis of the theory known as Mendelism, which by experiments on domesticated animals and plants has been proved to be a law and no mere theory. It is supposed by some authorities that it is these sports that give rise to new species in Nature.

Seasonal variation, as one might expect, is most marked in Indian butterflies. In any species that has two or more broods during the year, differences to a greater or less degree are to be found between the broods and, as a rule, the greater the local difference of season, the more apparent the influence on the wings of the butterfly. In some of the Satyrids the differences between the undersides of the two seasonal forms is startling and for many years they were regarded as different species: in the instances to which I am referring locality does not appear to affect the intensity of the dimorphism. In the

ISAAC KEHIMKAR

Common Mormon (Female, form *cyrus*) *Papilio polytes* – unlike the male, the female Common Mormon occurs in three different colour forms

dry season form the underside exactly resembles a dried leaf, while in the wet season form the underside is evenly striated and bears a row of eyes along the border. The reasons for seasonal variation are probably to be ascertained by a close investigation of the caterpillar stage and perhaps are due to the seasonal variation in the food plant; in some species it is possible that the characteristic has been inherited from bygone days and still remains though the original causes have disappeared.

Geographical variation presents a most interesting field for investigation and it is not unlikely that in geographical variation combined with Mendelism will be found the solution to the formation of species. As pointed out earlier, Nature is always urging a species to enlarge its sphere of action and, if a species spreads to a district which differs in climate or other particulars from its original home, it may, if it is a decadent species, fail to establish itself, but if it can contend with the change in the caterpillar food, the new enemies to be encountered with and the new climatic conditions, it will form a new colony. The different conditions may soon have an influence on the appearance, habits, etc. of the butterfly and a definite, easily distinguished, geographical race may become established. If the species is given to produce aberrations or become so by reason of the new conditions, as is quite probable, the sport, which is a recessive under Mendel's law, in the original home may become the dominant and, gradually swamping the normal form, establish a new species. If the habitat of the parent species and of the colonists is not separated by an impassable barrier, such as plains in the case of hill species, or hills in the case of plains species, or desert or sea, it is quite likely that the races will remain closely allied and can be graded in a long series. If, however, a barrier exists or becomes formed by geological changes the two races will gradually become more and more different as the centuries roll on, and should the changes in the earth surface ever bring them together again, they may be unable to interbreed and therefore must be regarded as species. Increase in elevation appears to have a considerable effect and I am not at all sure whether certain closely allied so-called species, found at different elevations, are not really conspecific, a certain feature being dominant in one area and recessive in another. Some of the inhabitants of the Himalayas differ on every watershed or in every large valley. There is a certain large Papilionid, *memnon* by name, that presents several remarkable features, not the least of which is that it possesses three forms of female, one of which is tailed, while the male and the other two females are tailless. Now in South India and Ceylon there flies an allied species called *polymnestor*, where the sexes are nearly alike, but are widely different from *memnon*, in that they bear a broad pale blue band above. Now *memnon* and *polymnestor* meet in the lowlands of Sikkim, but do not appear to trespass on each other's boundaries; yet certain known aberrations of *memnon* show a marked resemblance to *polymnestor*. I would not be at all surprised to hear that they were conspecific and that each is the dominant in its own area but the recessive in its neighbour's. In Nature it would seem that the recessive always becomes swamped by the dominant, unless it gets its chance under changed conditions, which are more favourable to the recessive than to the dominant, Breeding experiments on these lines might reveal all kinds of secrets. While speaking of the butterflies *memnon* and *polymnestor*, an interesting fact may be mentioned regarding the Andaman representative; the male shows a much closer resemblance to *polymnestor* than to *memnon* and is tailless, while the female appears to be a totally different insect and is almost exactly like the tailed female of *memnon*.

There are one or two other characteristics displayed by butterflies that I would like to draw attention to, as showing what an interesting subject their study can be. Butterflies possess peculiar instincts, regarding which we know very little. The latest theory regarding their eyesight is that not only do they fail to distinguish objects at a range exceeding twelve inches, but that they are also totally colour blind. Yet a butterfly can fly at a great pace unerringly through the closest jungle and appears to be able to distinguish his mate at great distances (the English "drinker moth" can, I understand, locate a female at a range exceeding a mile): he can locate flowers with apparently no scent quite easily and return to the same bit of carrion the next day, if he is so inclined; the female in selecting the correct food plant for her eggs proves herself to be a botanist of no mean calibre. Whatever a butterfly's eyesight may be, there is no doubt that he can spot at once the least movement that is out of the ordinary, as the collector will soon notice when stalking a wary species.

Many males display certain extraordinary features, known as secondary sexual characters; some have brands of specialised scales on the wings, others pouches and again others recumbent or erectile tufts of hairs; some species have a profusion of these features. In the Danaids the male can protrude long pencils of hairs from the end of its body. The functions of these characters are not understood and will only be solved by close observation work. Some butterflies, notably amongst the Morphids, emit a pleasing scent resembling vanilla, which remains for a long time after death. The primary sexual characters of the male are most extraordinary and have received a great deal of attention in recent years; their structure has in many cases proved to be of the greatest importance in classification, while in certain very closely allied species, an investigation of the genitalia is needed, before they can be accurately determined. Another curious characteristic is that displayed by the caterpillars of certain Lycaenids; they are attended by various species of ants, who in return for being allowed to suck the juices exuded from certain glands, look after the caterpillar and protect it from its enemies; the habit has got so strong that the caterpillar cannot live without its particular species of ant being there to protect it.

I will close this article, which already seems to me too long, with a few notes on where to look for butterflies and on the various localities that a collector may have an opportunity to visit. Flower gardens in general attract numbers of butterflies, but, generally, the rarer species are not to be found there. As a rule the collector must get into the jungle; beat the bushes and see what comes out; inspect flowers and especially flowering bushes most cautiously and carefully; don't neglect looking at carrion or manure; wet patches often attract butterflies in crowds; going along nallahs with running water is a paying business; males of rare species are often to be found on the tops of hills, especially towards midday. Don't try and catch a fast-flying large butterfly; you will only damage him; watch where he sits and then have a go at him; many butterflies, unless seriously disturbed, return again and again to the same spot and with adequate patience can be secured easily. Keep your eyes wide open as you walk along and if a butterfly gets up before you can catch him, chase him to see where he sits next and then stalk him warily; rapid movement is the one thing he can really see best and if he does not understand it, he runs away.

I may say that I have been collecting butterflies in India for 23 years and my interest increases as the years go on, my only regret being that I did not make better use of my earlier opportunities. I am a hard working individual at my profession and find that the pursuit of my hobby is the best rest possible from the cares of work, both during the periods I am at work and during the few spells of leave I have been able to obtain. It is a healthy out-door pursuit, with occupation for the evenings, and not only that, but I can look forward to plenty to do when I retire. I most strongly recommend anyone who has read through this article to take it up and can promise that the Society will give him or her every assistance; they have an excellent library, a good collection to refer to and experts, who will name anything. For those in the neighbourhood of Calcutta there is the Indian Museum to refer to and the Director of Zoological Survey will offer every assistance;

there is a most complete library and the very complete DeNiceville collection. Visitors to England should inspect the National collection in the basement of the South Kensington Museum, where Capt. Riley will do anything he can to help them. Don't forget that all the three institutions I have mentioned will be very glad to receive specimens especially from out-of-the-way localities.

Tytler's bushbrown *Mycalesis evansi*

[The collecting areas mentioned in the article are equally good for Butterfly photography which is a far superior pastime to collecting which we discourage on principle. Collection is also illegal as most species are protected by the Wildlife Protection Act 1972. In this day and age one should collect butterflies with a digital camera. — Editors]

■ ■ ■

The Panther As I Have Known Him

By Lieut Col A H E Mosse, F.Z.S. (Indian Army)
Year of publication: 1930

1. GENERAL

Whether he be called the Panther, as is usually the case in India, or the Leopard, the name by which he is generally known elsewhere, my subject is a beautiful and interesting animal who deserves a higher place on the Register of Indias Fauna he is sometimes accorded.

This is largely because he is overshadowed, in India by the Tiger, in Africa by the Lion. There are, however, various districts, especially in Western India, where tigers either do not exist or are few and far between, but where panthers are common enough. In such areas it is sometimes possible for the District Officer, when he can spare the time and will take the trouble, to cultivate the acquaintance of the panther in the ordinary course of his district touring and on occasional short holidays.

During periods of my service in the Political Department I have been so situated and venture to think that an account in some detail of the panther as I have known him, as also of methods of dealing with him, may be of interest and of some assistance to those who may have the opportunity and desire of making his closer acquaintance.

My experiences have been mainly in the northern part of the province of Gujarat in the Bombay Presidency, a country whose cultivated plains are bounded by a system of low jungle-clad or rocky hills increasing in height up to the borders of Rajputana. In the Mahi Kantha Agency, situated in the western portion of this area, many of these hills are covered with great boulders and, as will be seen, offer better opportunities of actually observing the panther than does country that is all under jungle.

Fine specimens of *Felis pardus* or *Panthera pardus,* as he is now called, are occasionally seen in North Gujarat and Kathiawar but, as a rule, the size attained in this part of India is, on the average, rather less than that of the large animal to be found in heavy forest country elsewhere. On the other hand the definitely small type, with very round head, which some sportsmen have sought to distinguish as a distinct species, is not found, except in immature specimens whose occipital ridge is undeveloped. This small type, which has been labelled 'leopard' as distinct from the larger 'panther,' does not ever, I believe, attain six feet in length, whereas I have never seen an adult male in Gujarat which did not substantially exceed this measurement.

Lieut Col A H E Mosse joined the Indian Army in 1897 and was attached to the 104th Wellesley's Rifles. In 1901 he volunteered his service for famine relief in the great famine and from that day he joined the Indian Political Service and served as Political Agent in the states of western India particularly in Bhavnagar. He retired in 1932 and was awarded C.I.E. for his services as Vice President of the Bhavnagar State Council. A great naturalist, he was particularly interested in entomology. He died at the age of 65.

Among some fifty panthers that I have personally measured, the length of an adult male in North Gujarat and Kathiawar has varied from six feet four to seven feet four and a half inches, with tails varying from twenty-six to thirty-four inches. My largest female was six feet four but I have seen one or two that were certainly larger. The largest males in these parts, of whose authentic measurements I am aware—though I have heard stories of larger ones—were two, both of seven feet eight, shot by that well-known shikari the late Lieut Col L.L. Fenton, one in the Danta State and one in the Gir Forest in Kathiawar.

The weight of a good male panther in Gujarat is something over 100 pounds. I have records of two only: one in Danta, measuring 6 feet 10½ inches weighed 114 pounds, ten hours after death: the second, a Kathiawar panther 7 feet 1 inch in length, weighed 123 pounds about eight hours after death. Both were in good condition.

With regard to coat and coloration, it may be observed that the fur is longer and the general effect somewhat greyer and darker in an immature specimen. Also that the ground colour as a rule tends to become paler in an old beast. Otherwise I have found no great

VIVEK SINHA

differences in colouring or pattern among panthers of North Gujarat and Kathiawar apart from some variation in the size and boldness of the rosette markings.

I have no evidence of the period of gestation; it is said to be about three months. There does not seem to be any particular season for mating. I have come across a mating pair more than once in the month of May I have not personally seen more than two cubs running with the mother but believe three is not an uncommon number.

In jungle country, one may occasionally have the luck of a casual meeting with a panther but, speaking generally, there are two methods of sighting a panther: 'beating' and 'sitting up'. So much has been written by competent authorities on the theory and practice of beating for tiger that I do not propose to say much on the subject.

The principles in beating for panther are the same, with these differences that, in the case of the smaller animal, such extensive areas cannot be beaten as in the case of the larger, nor is it of any use trying to beat a panther out of really thick jungle. A panther, moreover, especially a small one, will sometimes seek refuge in a tree and lie close while the beaters pass below him. A large panther is easier to beat than a small one but is not likely to travel as far when disturbed as a tiger. Provided these considerations are borne in mind, it is quite a mistake to suppose, as some do, that beating for a panther is always too uncertain to be worth while: in suitable conditions it can be done, as I have often proved. A point to which it is important to pay special attention is the placing of stops, a fact which many Indian shikaris fail to understand.

Apart from its being within reach of the spring of an enraged beast, too low a *machán* is a mistake owing to the greater risk of its occupant catching the eye of his quick-sighted quarry. At the same time it is essential to have freedom of movement and as clear a field of view as possible. It is therefore out of the question to try and conceal oneself as one can and must do when sitting up over a bait or 'kill'. Concealment during a beat is the

less necessary because, quick though the panther is to detect the least movement, it is remarkable how incapable he appears to be of identifying a motionless object. I have seen a panther, in a beat, come suddenly round a comer, catch sight of the *machán* on which I was sitting only twenty yards away and fully exposed, and stare straight at me for about ten seconds, then move quietly on, treating me as of no further interest. Other animals have this negative characteristic. I have known a sambur hind graze leisurely up to within fifteen yards of me as I sat with two shikaris on an open hillside. Then it became suspicious, but being to windward of us and detecting no human taint, still did no more than stare and stamp with its forefoot half-a-dozen times to try and make the doubtful objects, if alive declare themselves. With great wide open eyes she made a delightful picture. It was only when one of my men could contain himself no longer and sprang to his feet with a yell, that she fled, with the shock of her life. Provided, therefore, that you keep still and, if an animal stares at you, do not let its eyes meet yours, you are not likely to be detected. Failure to keep still may often lose you a chance. I have known a panther spot a movement in my *machán* at a distance of 400 yards.

The beat is generally a noisy affair with much shouting and beating of tomtoms, but in a small beat quieter methods are often better. During the period of waiting there is often much of interest to occupy one's attention in the various creatures that make their appearance, from deer and wild boar to jungle-fowl or a resplendent peacock. And, as the beaters approach, a troop of keen-eyed monkeys will perhaps give warning that their most hated foe is on the move. The blood tingles in your veins as you strain your eyes into the cover to try and distinguish that coat of brilliantly contrasting black and tawny-yellow whose hues yet blend in such amazing fashion with the light and shade of the jungle so as often, if motionless, to be quite indistinguishable twenty yards away. I suppose I have seen a hundred panthers in their native wilds, yet their wonderful gift of invisibility remains as great a marvel as ever.

Your panther may not appear at all. He has perhaps stolen away unobserved, between the very feet of the beaters or past an inattentive stop—or past you yourself as you looked in the wrong direction. But if he comes it will probably be sneaking stealthily along wherever there is any cover, breaking into a trot or gallop to cross an open space, or stopping every now and then to reconnoitre his front or listen to the uproar behind.

Before going on to deal with the second method of sighting a panther, that of 'sitting-up', it will be convenient to set down here some remarks on the character of the panther. Writers on big game have not always done him justice. Granted that, alongside the magnificent tiger, he is comparatively small beer; yet see him in an open glade in his native jungle, observe the muscular but agile symmetry of his form and the beauty of, his chequered coat in the rays of a declining sun; he deserves more than a second glance. Granted too, perhaps, that you may fairly, as one writer has done, describe the tiger as a gentleman, the panther as a bounder—though I question whether the difference in their respective characters is really such as to justify the distinction—but to say, with the same writer, that 'the panther is what he looks, a perfect swine' is a libel, that is, in the sense in which the term is used. For, come to think of it, it seems rather ridiculous that any sportsman who has really known that splendid beast, the fighting boar of India, should use the word swine on any animal as a term of contempt! To call the panther an arrant coward, as another writer has done, may be merely an instance of the folly of generalizing from a single case. No tiger hunter of experience will deny that the tiger himself can sometimes be a cur. Otherwise, to use the above-mentioned writer's own adjective, to label the panther a coward is arrant nonsense. The courage of the normal panther cannot be gainsaid.

Of course, the panther, like any other wild, animal, will seek to escape unobserved from the pursuit of men, whose superior powers he recognizes. Does not the tiger do the same? But wound him or get him in a corner and he is as ready to fight, and fight to some purpose, as is the tiger; perhaps more so. All said and done it is but his greater size that makes the tiger more to be feared. The panther displays at times a cool daring that the tiger will rarely rival. Make no mistake about it, he is a formidable foe and, if you begin by despising him, sooner or later he will give you cause to change your opinion and earn your respect. If you once grant him a title to respect you cannot call him a swine, much less a coward which he certainly is not.

In districts where they are common, panthers do much damage both to stock and to game, and no dog is safe in panther country. It may, therefore, be justifiable to write them down as a wanton killer. Yet, here let me tell a tale of a panther mother. I was once on the march between two camps. The bullock carts conveying my tents and baggage, etc., had gone ahead but had been delayed through getting stuck in a sandy river-bed, and I overtook them at a place where the road, or rather cart-track, passed for a mile or so through a patch of jungle. It was about 10 a.m. and I was riding leisurely along some twenty yards behind the rearmost cart on which was travelling a terrier who then owned me, with her family of three children about two months old. The cart was bumping along a track that was decidedly rough at this spot, and met a bigger bump than usual just as one of the pups had scrambled up on to a roll of tentage at the rear of the cart. The little chap lurched forward and fell to the ground yelping. He seemed hardly to have touched the ground, poor little beastie, when a yellowish streak flashed out from some bushes at the roadside and, before one could lift a hand, panther and victim had vanished.

That was not the end. As soon as I realized what had happened I turned and shouted to the sowar behind who was carrying my rifle. Then, looking round again, I was amazed to see the panther race back across the track between me and the cart, the pup still in her jaws, and regain the bushes from which she had just emerged, Why? There was no lack of cover on the other side. Investigation found the explanation in clear signs of the presence of a pair of cubs perhaps three months old—a family for whose feeding risks must be taken, but who were not to be left alone in the vicinity of danger. I sought vengeance for the poor wee pup, though without success. But I could not withhold my admiration for the combination of patient watching for and amazingly prompt seizure of an opportunity, the audacity of the successful rush, and the maternal devotion which took the risk a second time after the alarm was raised. Highway robber and dangerous killer perhaps, but a gallant beast to whom, mentally, I took off my hat. Would a tigress have dared that deed?

Generally speaking it may be laid down as an axiom that neither tiger nor panther will ever, unprovoked, attack mankind. It is the Jungle Law, by virtue of the respect for and dread of man in which all the jungle creatures are brought up. But when a panther takes to man-eating-fortunately not a very common occurrence-he is in some respects more dangerous than a tiger. He is definitely more audacious and will take greater risks. This is partly a matter of inherent character but also in part due to the fact that, from his habit of prowling round villages in search of stray dogs, etc., he has grown more familiar with mankind than is the tiger, with the result that he will not infrequently enter a hut in search of a victim, a thing the tiger will very rarely do. I have personally known half-a-dozen cases of people being dragged from their own huts at night by panthers who had been driven to man-eating by the effects, on game and cattle, of the famine conditions which prevailed in Gujarat after the great drought at the beginning of the 20th century.

So much for the character of my Subject. In the next part of this article I propose to discuss in some detail the theory and practice of circumventing him by 'sitting up'.

II. THE ART OF SITTING UP – ITS THEORY AND PRACTICE

By most Indian sportsmen the above heading will be at once understood. There may be others who will need the explanation that my subject is the method of sighting tiger and panther or leopard in India by 'sitting up' for the quarry in ambush over either a 'kill' or a live bait. To the stay-at-home reader the subject of tiger-sighting usually conjures up a vision of hunting him with a line of trained elephants in the high grass of the Nepal Terai, or else of driving it past with a horde of yelling beaters in the jungles of the Central Provinces. There are, however, places and circumstances in which neither of these methods is practicable and a good many tigers were killed in India by the silent and less spectacular method of sitting up. The majority of panthers shot were killed in this way. But, while much has been written about beating for tiger, the matter of sitting up is often passed over in a few words. An exposition, therefore, in some detail, of this form of sport, followed by an account of some illustrative incidents of personal experience may be of use to the novice who wishes to try his hand and of some interest to others.

Personally, when sitting up for panther, the difference to my success, since I have known what to do and what not to do, has been most marked. In my early days I was stationed for two and a half years in a district where panthers were common. During that period I bagged a number in beats but, as a result of sitting up about forty times, I never once saw a panther in daylight and on three occasions only had a shot at one after dark. Twenty years later, in the same district, I found that my panther gave me a shot on an average of rather better than one evening out of four and, in the great majority of cases, either in broad daylight or while it was still light enough to see my foresight clearly. It was not that the panthers of the later generation were more numerous or less sophisticated than their forebears, but that I had learned from the lessons of failure. General Wardrop–*vide* his *Days and Nights with Indian Big Game*–has had a very similar experience. These are facts which speak for themselves.

There may be something repugnant in the idea of sitting over a live goat and watching the unfortunate creature done to death but after all it makes no difference to the goat whether it meets its fate under a human eye or not, and the practice is universal in India of tying up a live bait in order to obtain a kill, whether with the object of sitting over the latter the following day or of beating the tiger out of the adjacent cover in which it lies up after its meal. Moreover, when you sit yourself over a live goat, the latter has a greater chance for its life than in the former case, since it should be your object, if opportunity occurs, to shoot the panther before it seizes its prey and, in practice, it is often possible so to save the life of the goat: in such event it is but fair to grant the individual goat immunity from having to run the risk a second time. It is quite a mistake to suppose that the goat has any anticipation of its possible fate, while death, if it does come, is speedy and, be it borne in mind, the sacrifice of one goat and the resultant death of its slayer means the saving alive of many, for the depredations of panthers on the village herds are continual.

A great point, moreover, in favour of sitting up, which is not appreciated by those who have not given it a fair trial, consists in the opportunities it affords to the lover of Nature of attaining a greater intimacy with his Mistress and of being initiated into some of the secrets of the jungle. Sitting up, too, is an excellent discipline. It is no game for the man who cannot possess his soul in patience and may be termed monotonous by those who have no thoughts beyond the actual killing of their quarry. For me it has a great charm, as it must always have for those who love the jungle or have in them anything of the naturalist. There is a fascination which never palls, especially if you are in the heart of the jungle, in that silent watch and ward as you listen to the gradual awakening of the voices of the night around you. And never in the daylight can one experience quite the same kind of thrill as when one becomes conscious of the near presence of a wild beast, invisible in the gloom of night.

Like many other wild animals both tigers and panthers will often fail to notice or identify a motionless object, but their eyes are marvellously quick to detect the least movement around them. Their ears are equally sensitive. They are also by nature extremely wary and suspicious. To an insufficient appreciation of these characteristics is due the failure of many an unrewarded vigil.

It follows that in making one's preparations it should be one's aim to avoid anything likely to arouse suspicion and to make oneself as invisible as possible. To deal first with sitting up for a panther over a live bait:- The mere fact of a single goat being tied up alone may be sufficient to arouse the suspicions of a panther of experience, but will not as a rule deter him from attacking if, after studying the situation, he detects no sign of the presence of man. To begin with: I assume that a panther has been approximately located, probably by his tracks–it may or may not be after a kill–in some particular hill or patch of jungle. The first thing to be done is to select a site for the *machán* in a quiet spot within a moderate distance of the beast's retreat–not so near as to run a risk of disturbing him, but near enough for the goat's calling to be audible in the proximity of his lair. The most usual site for a *machán is* a leafy tree, though a sheltered rock or a thick bush with a bank behind *will* sometimes afford an excellent position. A platform of stout branches is made in the tree or–this is what I have generally used–a native *charpoy* may be slung up and securely fastened by ropes among the branches, with a native quilt or two and a couple of cushions, or a round stool to sit upon and give your legs freedom. This may sound *sybaritic* but *is* merely common sense. *You must* be able to keep still or, if you have to make an occasional movement, to do so without noise. This is impossible if you are not comfortable.

The foliage of the tree, alone, will rarely, especially in the hot weather, provide adequate concealment. It is therefore necessary to screen yourself all around with branches. An effective method of doing this is by means of light hurdles made of leafy branches, preferably of the same kind of tree as that which holds the *machán*. Such hurdles are best constructed at a distance so as to take less time in the actual preparation of the *machán* on the spot. In the front screen facing the goat a loophole is constructed, with a firm cross-bar at a convenient height on which to rest your equipment.

If you are anxious to get your shot before dark, it is sometimes advisable to have another loophole or two for use in one or both of the side screens to command a probable line of approach.

This last provision for a good view, however, must not be overdone. I am convinced that the most frequent cause of failure is inadequate concealment of oneself. Colonel Stockley writes of 'so arranging things that you have a good view all round' but it is significant that he has had little success in sitting up. I used once to be of the same opinion–in those days the panther kept away! As a rule it is best to be content with a loophole in front only, with perhaps a small peephole on one side.

One always knew in theory that a panther was a wary beast and quick of sight and hearing, but the superlative quality of his watchfulness was not really brought home to me until I had had opportunities, watching in my turn, of observing it for myself. I have seen a panther sit or lie watching his 'kill' or a tied up goat for more than an hour at a time. I have seen one sit staring at my *machán* –I knew it was not a good one, nor well placed–from a distance of eighty yards and then slip quietly away never to return. On many other blank evenings that is doubtless what has happened, the panther has taken observations, himself unseen.

No doubt a good field of view might sometimes enable you to see the panther earlier and give you a daylight shot, but it is not worth the risk of his spotting some slight movement from one direction, while you are gazing in another.

In this connection the matter of height of your *machán* has to be considered. The higher you are, the less likely to be seen. At the same time, especially at night, accurate sighting is definitely not as easy from a height as on the level. In my opinion about fifteen feet above ground is best. I do not like a *machán* above twenty feet.

It is important to make sure that there is no background of clear sky behind your head: a slight unavoidable movement in shadow may pass unobserved, even by a panther's keen eye, which against the sky would be noticed at once. It is a wise precaution to place a man in the *machán* when it is ready and study appearances yourself from outside.

Another matter, which may make all the difference is the location of your tree relative to the beast's probable line of approach and the position of the bait or kill. When possible your tree should not be an isolated one; if there are others around it is less likely to invite a beast's particular attention. It may sometimes be impossible to judge from which direction a tiger or panther will come and in any case he will often prowl all round before coming near. But, especially if the *machán* be at the base of a hill in which he is known to have lain up, one may often judge correctly his most probable line of advance. In such a case it is advisable that the *machán* should not be situated in his direct line of sight as he approaches but, if possible, to one side. Though making observations all around, his attention is directed mainly towards the bait or kill; therefore anything suspicious about a *machán* beyond the bait but in the same general line of sight is more likely to attract his notice than if the position be looking towards the bait from one side. If your *machán* is between the panther and the bait you run the risk of his approaching from behind you unseen and unheard and possibly halting beneath or near your tree where he is liable to hear your slightest movement. I have known a panther to sit exactly beneath my *machán* and depart without my having had any suspicion of his presence until examination of his tracks afterwards told what had happened!

The distance of the *machán* from bait or 'kill' is another factor of importance. The greater the distance the less chance of your being detected. If you do not mean to sit after the light has faded forty yards is not too far. But by good moonlight or when using a flashlight torch twenty to twenty-five yards is the best distance.

There are times when no suitable tree is available. If there are high bushes to afford cover a platform on four poles may be an adequate substitute. Otherwise, in the absence of a tree, an efficient hide can often be made among bushes or on the bank of a nullah. In such a case it is as well to arrange so that the beast cannot easily get at you in the event of things not going quite right!

With regard to your conduct of affairs when in the *machán* it should be unnecessary to say that patience and quiet watchfulness are essential. If you have to move–and there are times when you must–your movements must be deliberate and of the slowest. If you propose to sit late and have food with you, see that it is not wrapped in paper that will rustle, and avoid tins. As General Wardrop most truly writes 'the sound of metal on metal is the danger signal of the jungle,' so be most careful of your equipment such as whistle— anything of metal. I have occasionally known a signet ring on my finger to knock against a rifle barrel, so now I always make a point of removing it.

You must be careful to avoid having any dry leaves on the *machán* close to your loophole or in any place where there is a risk of your touching them and causing them to rustle. Do not have your loophole too large, but see that its edges are fairly clear cut; a projecting leaf or twig which may be of no importance in daylight can become a serious impediment to clear vision in the dusk.

While dry leaves in the *machán* might betray you, they may on the ground betray the approach of the panther. There are, therefore, occasions when they can with advantage be spread at the foot of your tree.

Practically all that I have said so far applies in equal measure whether you are sitting up over a goat or a 'kill'. But there are further special considerations to be taken into account in the case of a 'kill'. The natural instinct of the wild beast would seem to lead it to apprehend the possibility of an enemy having found its kill and seized the opportunity of lying in wait for its return. Hence, presumably, the caution usually displayed in returning even to a natural 'kill', by which I mean a kill obtained in the ordinary course of hunting and not of an animal tied up as a bait. This caution is intensified by the suspicions aroused by a tied-up bait especially when, as is often the case, the tiger or panther has learned from previous experience the risk of danger associated with such.

It follows that a natural kill is the best. But this is not usually to be found just when YOU want it. And when one does occur, the corpse of the victim is by no means always left in a place commanded by a suitable site for a *machán*. Moving a 'kill' is not to be recommended.

Apart from the fact that the moving of a 'kill' engenders suspicion, the powers of scent of the great cats are poor. A panther has been seen to scent up the drag of a 'kill' that has been moved, but you cannot count upon his doing so and I doubt if a tiger will ever do it. If, therefore, it be absolutely necessary to do so at all, do not move a 'kill' more than a few yards and then not out of sight of its original resting-place.

If then you want a 'kill' to sit over-and failing a natural one–you must first choose your *machán* site in a likely place and tie up your bait securely so that it cannot be dragged away. Presumably in order to take advantage of the better field of view, both tiger and panther habitually perambulate along the jungle paths. The junction of two paths or nullahs or of a path and a nullah is therefore a good place.

You may have baits tied up at several spots. It is worth while having a *machán* made beforehand at the likeliest of these. You will then avoid the risk of the beast being disturbed by the noise of the erection of the *machán*. When you have to put up your *machán* after the kill has been made, and there is a likelihood of the beast being within hearing, it is sometimes a good plan to send a couple of men moving in the direction in which he is believed to be, the object being to disturb him sufficiently without scaring him, in order to make him move out of hearing. But this requires to be done with judgement,

So far as possible there should be nothing suspicious about the *machán,* to catch the panther's eye, which was not there when he made his kill. The work of constructing the *machán* after a kill should, if practicable, be finished by 2 p.m. You can never be sure when a tiger or panther will return to his kill. It depends upon place and circumstances, upon the idiosyncracies of the particular beast and the state of his appetite. When the 'kill' is in a secluded place and he has lain up near by he may return at any, hour, but as a general rule it is advisable to be settled down in your *machán* by 4 p.m. in the cold, and 5 p.m. in the hot weather, at latest.

In my experience it is rarely worth while sitting up for a panther for much more than an hour after dark. If not suspicious of danger he will often appear before the sun goes down, but the most likely time is the hour immediately after sunset. My experience of tigers is limited. I have shot one tigress over a kill in bright sunshine, but as a rule they are later than panthers, and, if it is not too cold, you should be prepared for an all night vigil. General Wardrop says that if a tiger does not arrive by 9 p.m. he is most likely to turn up about midnight or at dawn.

A good bait for a male panther is a young male buffalo from six months to a year old. A goat may be completely devoured or not enough left for a second meal. A buffalo kill in my experience usually ensures a return if nothing has made the panther suspicious.

If you intend to sit after dark the question of light becomes of importance. If the moon be near the full, well and good; but even then moonlight can be very deceptive.

A good electric lamp makes all the difference. There are several devices of the kind on the market. One is a lamp with a shade intended to be hung above the 'kill' and throw a light directly down upon it; a controlling wire is connected to the *machán*. I have not tried this, but it seems to me a weak point that a fixed area only can be lighted up, while it may not always be possible to fix it in the right position above the 'kill'. A better method is provided by a lamp fixed in the *machán* with movable bull's eye which can be adjusted to throw its beam in the required direction. This is the type of lamp advised by General Wardrop who has used it with great success.

The only kind to which I have given a fair trial–and found very successful-is a powerful cylindrical spotlight focusing torch which is clamped to one's gun-barrel so that gun and light are together aligned upon one's target. This torch throws a beam of light which effectively light, up the 'kill' at a distance of twenty to twenty-five yards. It also has the great advantage of, at the same time, illuminating one's foresight. If the light be suddenly thrown directly into an animal's eye it may frighten him off. If, however, the light strikes the eyes from the side he will usually look up and stare, giving you plenty of time for a shot. If the light does not strike his eyes the beast will in my experience take no notice whatever.

There are some who consider sitting up by daylight permissible, but would bar the use of an electric light at night as 'unsporting'. So far as I understand, the objection is that it makes the business too easy.

In any case the objection may be dismissed in a word: it is merely absurd to suggest that it is easier to shoot with an electric lamp than in broad daylight. The second objection is a little more difficult to deal with. An inadequate illumination is hardly better than none at all. But there are several makes of suitable and inexpensive lamps to be purchased nowadays which possess ample power: it is only necessary to be careful that the battery does not get exhausted; wherefore in a sporting trip a couple of spare batteries should be carried. And it is reasonable to assume that no intelligent sportsman will go to his *machán* without testing the capabilities of the lamp he intends to use.

I have dealt at some length with the theory and practice of the art of sitting up, but it has seemed worth while to discuss the subject in detail. In subsequent parts I propose to illustrate what I have already said by describing some personal experiences of the panther and his ways–both by night and day.

III. SITTING UP EXPERIENCES

In giving some account of personal experiences I shall confine myself to those which present definite points of interest or are otherwise illustrative, either of habits of the panther which I have already described, or of lessons to be drawn by the observer.

On one occasion, sitting in a low and unavoidably conspicuous *machán* at the foot of a hill near a village called Chelana, in the little State of Sudasna in the Mahi Kantha Agency, I suddenly observed a panther sitting on his haunches, looking at the *machán*, about eighty yards away. He sat for five minutes, then disappeared for good. In this case he may have spotted a movement of mine before I saw him or merely did not like the look of the *machán*. The mistake here had lain in 'chancing' a bad *machán* owing to unwillingness to move the kill—a natural one. It is undoubtedly wiser not to move a kill if this can be avoided, but some panthers will follow the drag of a kill moved for a little distance and it is better to risk their not doing so rather than to occupy a position foredoomed to failure. In the case of an old hand your only chance, as a rule, is a natural 'kill' within sight of a site for an inconspicuous *machán* and out of sight and hearing of his daytime retreat.

On one occasion at Chelana I sat up for a pair of panthers, supposed to be a small male and a female. One appeared on the rocks above me an hour before sunset, the other ten minutes later. Through the glasses I came to the conclusion that they were a mother and big male cub, very nearly as big as herself. In such circumstances the average panther affects, for most of the time until he decides on action, an air of complete indifference to the tied-up bait below. In this case, however, it was noteworthy how the attention of the cub was concentrated on my goat, off which he could hardly take his eyes. At length he went up to his parent and put his nose against her cheek. She licked his head and then he left her and moved down to a lower rock. It was exactly as if he had asked permission to try and tackle this business by himself and she had bid him go, with her blessing–and a warning to go slow and not rush things. From the lower rock he watched the goat intently for five minutes, then slipped down into the bushes below. Immediately afterwards the mother climbed a small tree that grew alongside the rock on which she had been sitting, apparently in order to obtain a better view.

Some ten minutes later the cub came into my sight again through a peephole, creeping very slowly and with the utmost circumspection. About ten yards distant from the goat, who did not see him–he was partially concealed by a small bush and some grass—he lay on his stomach, motionless, for five minutes, then he moved forward at a crawl, inches at a time and with frequent pauses. He took not less than fifteen minutes to cover ten feet of ground: all the while his tail never ceased twitching, while his eyes remained fixed with the utmost intentness on his quarry. One could sense the conflict between the impatience of his youth and his determination to obey parental instructions and go slow, his whole body tense yet quivering with excitement. It was fascinating to watch him. At length he was only fifteen feet away, now fully visible through my loophole, while his tail whipped from side to side. One part of me would have liked to wait and see what sort of a job he made of the attack, but that would not have been fair to the poor little goat, still all unsuspecting, and I made an end. It had been most interesting.

The following evening I tried to tempt the mother out but without success. I heard her, however, quite near, calling for her lost son, a peculiar murmuring sort of note which gradually changed into the well-known 'sawing'. This 'sawing' is, I think, usually uttered as a call to a mate, or, as in this case, a cub. I saw more of this lady not long after when she had taken to herself a husband. I had had a *pada* tied up for a possible tiger near a good tree well out in the open, some 150 yards from the bushes at the base of the Chelana hill, and it was killed by this pair of panthers. I sat up an hour before sunset but, in view of the position of the 'kill', did not expect to see anything of the panthers before dusk. However, after half-an-hour I heard them–there is no reticence about a panther's courting–and they shortly appeared out in the open. For the next three-quarters of an hour I was treated to an exhibition of the panther's married life. The lady was not exactly backward in coming forward–in fact, I never saw a more brazen hussy!

The daylight was beginning to fail when they approached my tree and the female came and stood over the 'kill'. Just then the male, who was under my tree, gave a sort of whinnying call; she at once turned tail and joined him and they retired to the shelter of the bushes at the foot of the hill. I wondered if my presence had been detected, but next moment the explanation came with the loud moaning cry to which a hyaena sometimes gives vent when in the neighbourhood of a panther's 'kill' and he knows the owner is near by. Presently I saw him fifty yards away. Twice he made a circle round the 'kill', keeping up his peculiar moaning. At length he came up to the 'kill' and started his meal, but did not really settle down to it, being obviously uneasy.

By this time the daylight had gone, but the moon was near the full and its light was bright. The hyaena had been at the 'kill' for about five minutes when the panthers made their appearance again, strolling leisurely up, and both lay down some thirty feet away, watching the hyaena but, apparently, making no hostile demonstration. The hyaena was clearly worried: he moved a few feet away from the 'kill', began his moaning again, and then, after a couple of minutes' hesitation, retired from the field. After he had been gone a few minutes the male panther walked up to the 'kill'. It had been an evening that afforded, an exceedingly interesting example of the manner in which sitting up may provide one with fascinating peeps into the intimacies of jungle life.

A friend of mine once witnessed a fight between a small panther and a hyaena, over a 'kill', in which the latter had the best of it, and there seems some reason to believe that the panther holds the hyaena, with his powerful jaws, in considerable respect. But the respect is mutual. The hyaena ordinarily is a coward and I have seen one decamp with the utmost celerity at the sound of a growl from a panther returning to his 'kill'. In Somaliland I have seen a male panther give way on the approach of a couple of hyaenas but the odds were heavily against him, as in a few minutes there were half-a-dozen hyaenas on the 'kill'-and this was the big spotted hyaena of Africa. In my experience the striped hyaena of India is very chary of settling down to a 'kill' if he knows the panther to be anywhere about.

Once mated, a pair of panthers will keep together for some time, but if she lose her mate in the early days the female is not long in consoling herself.

VIVEK SINHA

When a pair of adult panthers is working together it is almost invariably—I have seen but one exception to the rule–the female who makes the 'kill', unless the quarry be beyond her powers. I have never known a female kill the half-grown buffalo of which a male will make short work. As says Kipling, 'the female of the species is deadlier than the male', or it may be that the male is a believer in the maxim 'let the women do the work'!

Another case of a mated pair which the attendant incidents make worth recording occurred in a jungle in the Idar State. I had been sitting over a goat for an hour listening to the row the panthers were making close by when the female, who had not previously showed herself, rushed the goat, seizing it as usual by the throat. The goat was a fairly big one and for a few seconds there was a struggle, during which I could not make sure of my shot. Then, when things had quieted down, the goat was apparently done for and there was no hope of saving it, so I withheld my fire, hoping the male panther would appear. The spectacle was unusual, the panther, not a large female, was sitting on her haunches holding the seemingly dead goat on its feet in a standing position, maintaining a tight grip on the throat. As the body of the goat swayed against her she pushed it gently away with a fore—paw. This happened two or three times. For, I should think, two minutes, she remained like this, then relaxed her grip and the goat collapsed. She surveyed the body for a few moments, then seized it by the head and attempted to drag it away into cover. In this she failed; the goat was firmly tied by a rope to a stump. Then she pulled at a hind leg, without success. Again she tried another leg. This time it seemed to me that the rope showed signs of slipping off the stump. I decided to take no risk and fired. She dropped dead and I called up my men. Just as they arrived the 'dead' goat gave a kick and stood up! On examining it I found that, although it had recovered from the temporary strangulation, it was undoubtedly badly wounded; so I put the poor thing out of its pain.

This incident illustrates how the panther ordinarily kills a goat or similar animal by strangulation, rather than by the wounds it inflicts, though these would no doubt as a rule eventually prove fatal. It is perhaps this habit of holding its victim's throat in a strangle grip for a minute or two, or even longer, which has given rise to the common belief that the first thing a panther does, after killing, is to drink the blood of its victim from the throat. I believe this idea to be entirely erroneous. My reasons are, in the first place, that if you examine a goat killed by a panther you will ordinarily find only the two punctures on each side of the throat. During the continuance of the grip these puncture holes are occupied by the panther's fangs and no blood could flow through them. In order to draw out the blood the grip would have to be shifted, in which case there should be other fang marks, but of such you will, as a rule, find no sign. Again, if the panther were to suck the blood while gripping the throat, one would, while watching the performance, as I have done, observe the motions of its own throat. But one sees nothing of the kind; there is no sign of any motion of sucking or swallowing while the panther maintains the grip. A final conclusive argument, to my mind, lies in the absence of blood on the throat of a panther's kill, or on the ground below. If a panther sucked blood from the throat, the action would necessarily induce a flow of blood from the wounds which would continue for a time after he had released his hold. But, in actual fact, it is remarkable how rarely one sees any blood at all. It is likely enough that a beast may lap up some of the blood which he finds pouring out when he tears open the stomach—I have known a lion do so but that is another matter. The story one hears now and then of a panther—or tiger for that matter—who has left a kill uneaten having contented himself with drinking the blood is, I am convinced, based on a delusion.

Another panther, in Kathiawar, which escaped after being hit in the shoulder, was an interesting beast. I was sitting on a sheltered rock overlooking the 'kill,' a goat, in the

nullah below. The 'kill' was securely tied, by the neck, to a peg. It was a dark and windy night and, after dark, I had to rely entirely on my ears. After a time I imagined once or twice that I heard something resembling the sound of eating, but was not sure. At length I turned on my light and there was the panther on the 'kill'. The battery of my torch had run down a little and I think I rather misjudged his position and hit him somewhat farther forward than I had intended. At the shot he disappeared. On investigation I found quite a lot of blood on the spot, with a piece of bone an inch long, presumably from the shoulder. There was again nothing to do but leave him till the morning. But a very interesting feature was the condition of the goat. Now a large panther, when his 'kill' is a fairly big animal such as a young buffalo, will often start his meal at the root of the tail, otherwise he invariably commences at the stomach, continuing with the flesh off the ribs, and eating the whole of the body before touching the neck which he often leaves. In the present case the panther had eaten very little the previous day and that from

the stomach. Now, to my surprise, I found that the stomach had not been touched, but that he had been working all round the neck, which was nearly cut through. He must have been at it for five or ten minutes before I turned on my light, none too soon. I see no reason to doubt the correctness of the obvious explanation that, when he found his kill fastened down by the neck so that he could not drag it away, he deliberately set to work to cut off the head. The incident sheds an interesting light on the panther's intelligence. I have never heard of a similar case and one has always been inclined to look upon it as a sign of their lack of intelligence in some respects that neither tiger nor panther has ever been known, so far as I am aware, to cut with his teeth, as he could so easily do, the rope which often prevents him from dragging his kill away. The present case, however, certainly appears to indicate the exercise of something akin to definite reasoning. Later on I shall have a story of tell of a different kind of display of intelligence by a panther during a beat.

While neither tiger nor panther will cut the rope which fastens his 'kill', he will pull until he breaks it by sheer strength if he can. And I have known a panther to carry a goat away, and on another occasion to remove a 'kill', in each case breaking the rope with which it was tied by the sheer force of his initial rush.

A panther returning to his kill is usually the personification of caution. I have seen one circle all round three or four times listening and peering intently in every direction. The instances I have already cited show how patient he can be before attacking a goat, and his approach is usually silence itself. But if he is hungry and unsuspicious he will at times be amazingly impetuous. I once sat up over a goat in a jungle at the foot of the Girnar Hill in Kathiawar. The goat was calling loudly and I had not been in position more than a few minutes when I heard the noise of some heavy animal plunging through the jungle. It might be an alarmed sambur but sounded more like a buffalo. Straight towards me came the creature, whatever it was, and, to my amazement, a large panther broke cover and pounced upon the goat.

On another occasion a panther was seen in the morning on some rocks close in front of which was an excellent tree in which I had a *machán* constructed. At 4 p.m. the panther was out on the rocks again and I succeeded in approaching behind the tree and climbing up unnoticed. Then I had a flock of goats driven past me and one of their number tied up as they halted beneath me. The panther immediately sat up and took notice. The flock passed on. Within five minutes the panther had slipped down and in five more was on the goat, at 4.30 on a hot May afternoon. I shall never forget the expression of the gleam in that beast's green eyes, as it sat up facing me, its fangs in its victim's throat—it was the acme of gloating ferocity, giving a remarkable impression of cruel joy in the act of killing.

The great *carnivora* kill for food and it is no doubt perfectly true -that they do not ordinarily kill at one time more than will satisfy their needs. But there is equally no doubt in my mind that, given the opportunity, they will now and then slay for sheer lust of killing. I have known several cases of both tiger and panther making two and even three kills at a time and not attempting to dispose of more than one.

The panther is an extremely efficient hunter, but, like the lion in Africa, he has no hesitation in partaking of a ready-made meal. I met with an interesting instance of this in the little State of Vijayanagar. I was sitting one evening over the kill of a tigress. My *machán* was in a tree in a nullah, the banks of the nullah to my right and left front were almost as high as the level of my eye, so the position was not a good one, but the best available. A little before sunset I suddenly observed a large panther sitting on his haunches on the left bank, looking down at the 'kill'. I was not prepared for sitting up after dark and thought my chances of getting the tigress in the circumstances were small, while the panther was a fine specimen; so I took the shot, an easy one, only to make one of those inexplicable misses that will occur now and again to every shikari. One likes to assume they are due to faulty cartridges, and no doubt they sometimes are, but not always! I had not much hope of anything after that, but decided to sit on until dark. Presently a mongoose appeared, one of the dark grey species with black tip to his tail, the Ruddy Mongoose *Herpestes smithi* which is not uncommon in these hills, and after some hesitation settled down to a meal. The sun set and the light was beginning to fail when, from beneath my *machán,* came the sound of a stone turned over by some passing foot. I looked down and there was another panther, a good male but, I thought, definitely smaller than the first, creeping stealthily forwards towards the kill. The mongoose saw him too and gave his peculiar little alarm cry. I expected him at once to turn tail, as I have seen them do on other similar occasions. But, to my amazement, he sat up on a stone and proceeded to hurl abuse at the panther who took not the slightest notice but continued a slow stealthy approach.

No doubt the mongoose considered himself entitled to his perquisites from any 'kill' pending the arrival of the rightful owner to whom he could not object to give way. But he did strongly object to trespassing by a beast who was well able to kill big game for

himself. And so the little thief sat there, pouring out upon the head of the big thief the vials of his wrath in purest Billingsgate. It was fascinating to watch the little beggar's audacity; the panther was now within six feet of him. I wished afterwards that I had waited to see the end of the little drama. But the light was going and the shot offered was just what I wanted; so the bigger thief paid the penalty for his meditated crime. Subsequent investigation of tracks showed that this was, as I had thought, a different panther and that the first had got away unscathed.

The mongoose, of two species–the common kind *Herpestes edwardsi* and the larger darker one with a black tip to his tail–is a frequent visitor to a 'kill'. Sometimes there may be a party of two or three, when their bickerings over a carcase are amusing to watch. They are usually very nervous and on the alert and their peculiar alarm call and hurried departure will often give the first warning of the approach of tiger or panther. Once a big mongoose was busy on a tiger's 'kill' when a grey jungle fowl came up pecking about on the ground. At one time she was within two feet of the mongoose but took no notice of

VIVEK SINHA

him whatever, nor he of her, although the mongoose is a deadly foe of poultry. It was an interesting illustration of the fact, which has often been noted, that the denizens of the wild, whether by means of telepathy or some form of instinct, appear to possess a faculty of divining an absence of hostile intent on the part of a hereditary enemy, and in such cases will show no fear of him.

Other visitors to a 'kill' are hyaenas and jackals, the latter always very nervous. I once heard of a case of a porcupine making a hearty meal inside a dead buffalo. Of birds, the tree pie sometimes joins in the feast and I have known a myna to do so. Then of course there are the vultures and crows and occasionally an eagle of plebeian tastes. I have seen a jackal watch a concourse of vultures in the air, collecting over a kill, and take them as guides to the spot. The celerity with which vultures will dispose of a dead animal must be seen to be realized; it is a gruesome sight. After the pair of panthers mentioned above had been skinned and thrown out the two skeletons were picked clean in fifteen minutes.

I have never seen the *chaus*—the common jungle cat—on carrion. But I have seen one, and another time the grey spotted desert cat, take great interest in a tied-up goat. I do not think I know of any expression more coldly cruel than that of the pale green eyes of the jungle cat. It was when sitting over a goat in the Danta State that I once, at dusk, observed a specimen of that rarely seen animal, the ratel or honey-badger, stroll past within a few yards.

The panther sawing note is usually, I think, though perhaps not invariably, a call to a companion. His growl of menace is familiar but a pair together sometimes give vent to weird sounds that are not easy to describe. Once, on the borders of the Gir, I heard a dying panther give vent to a roar of surprising volume. There was no mistake as to whence the sound came, otherwise I should have put the roar down to one of a party of lions which I had heard, not far away, half-an-hour before, and which actually killed a buffalo only a quarter of a mile away from me while I was waiting for the panther. I heard the hubbub as they were driven off by a crowd of villagers but did not at the time know what had happened, or I should have tried to be in wait to see something of them, not to shoot–the lions of the Gir are strictly protected, a fact which has led to a considerable increase in their numbers of recent years.

I have referred to instances of a panther breaking a rope and carrying off a goat with his initial rush and I have known of one dragging a good sized donkey fully three hundred yards over rough and hilly ground. But the most remarkable display of sheer strength by a panther which I have come across was that of one—from his pugs a not particularly large male–which killed a big cow, a fine specimen in good condition, of the large Kankrej breed. The point of interest lay, not in the mere killing of the cow, but in the fact that, seized by the throat as she lay, she was so firmly pinned down that she was absolutely unable to move, as evidenced by the fact that she was found dead in the ordinary recumbent position of a cow at rest chewing the cud; there were no signs of a struggle and no wounds beyond the usual fang punctures in the throat; the panther had been disturbed before he had commenced his meal.

An unwounded panther will rarely charge unless cornered; I can however give one instance of an apparently unprovoked attack. One afternoon two men, Thakardas by caste, came into my camp, one a youngster of eighteen or so, rather badly mauled about the right side and arm. Their story was that they were occupied in cutting wood, at about mid-day, when a panther suddenly sprang out upon them, and seized the boy by the right side, below the armpit. He said that he then seized one of the panther's ears with his right hand, and with his left its lower jaw, while the other (his elder brother) sprang astride the panther's back and smashed its skull with a blow or two of a heavy stick! It

sounded a tall yarn, but they were so matter-of-fact about it that I have no doubt it was quite true. I sent out a couple of men with the uninjured brother and they brought in the dead panther—a rather small female—with a broken skull, but no other sign of injury. This attack was a remarkable one, but probably to be accounted for in one of two ways. The panther may have had a young family near by (though no trace was found of any), or she may have been in some manner taken by surprise and thought she was attacked.

I once witnessed an example of the speed of a charging panther which afforded as narrow an escape from a 'nasty mess' as I have any desire to see. I had a shot on foot at a panther bounding away at about thirty yards distance and hit him, as it appeared, in the hind quarters, the bullet raking forward. He fell and lay struggling. From his movements I thought his spine was injured and that he was therefore helpless, so fired my second barrel at his head to finish him off. Just as I pressed the trigger he made a snap at his wounded quarter and I missed. At the sound of the shot he looked up, and in a moment, for his spine was not injured, was charging straight for the first man he saw. This was a Rajput shikari who had come up unknown to me, and was standing where he had no business to be, about twelve yards to my left. I dropped my rifle and snatched another from the orderly at my side, but before it reached my shoulder that raging living thunderbolt was on the shikari and down went man and beast together. Then occurred a remarkable incident. The shikari was wearing on his head a *pagri* consisting of many folds of a mixture of silk and cotton, a fine but very tough material; this was not merely wound round his head but was in some way fastened to his hair. As he fell the panther struck with both paws at his head. Result, that the next instant the spectacle was provided of the man on his back and the panther on top, struggling, for the moment unsuccessfully, to release its claws entangled in the *pagri's* clinging folds! It was ludicrous if one had leisure and inclination to laugh. There was just time however to run up close and put in a finishing shot, before the panther, not a large one, could do any damage. The dead beast was pulled off, and the shikari picked himself up quite unperturbed, true Rajput that he was, and none the worse except for a single scratch on the back of his hand. of course he would not have escaped so lightly but for the fact that the panther's first wound was a mortal one and the effort of the charge had absorbed most of its failing strength. Nevertheless had that first blow been three or four inches lower, the whole front of the man's face would probably have been torn down by those deadly claws.

Though essentially a nocturnal animal, the panther now and then begins his hunting some time before sunset or continues it after sunrise. But you will very rarely meet him. With his quickness of sight and hearing he will invariably detect you first and unless he chooses to be seen you will not see him. One of the rare occasions on which I have met a panther casually in the daytime when not looking for him is worth recalling. It was near Himatnagar in the State of Idar. I was encamped there shortly after the installation, at which I had assisted, of the then Maharajah, that grand old Rajput sportsman Sir Pratap Singh. On the river close by was a charming waterfall, the face of which, wherever the flow was not too strong, was covered with maiden-hair fern. At the foot of the fall was a deep pool perhaps twenty yards in length, which at that time was inhabited by a small mugger and was known to us accordingly as the Mugger Pool. I never saw more than the nose or top of the head of this *mugger* but judged him to be not more than three or four feet long, and therefore no obstacle to a swim. The delight of those early mornings in the hot weather! A plunge in that beautiful cool water, a swim round, and then to clamber half out and sit amidst the delicate green of the maiden-hair beneath and surrounded by the silver cascades of the most perfect natural shower-bath; what mattered the burning hours of the day to follow, grilling in an office tent–the present was bliss! It was on such a morning that I was striking across the pool, heedless of the *mugger's* rights, when I happened to glance up towards the rocks above on the far side, and gasped! For there, some fifty or sixty yards above me, on his haunches sat a panther surveying with interest this new white *mugger* or whatever it was that the water had produced. I wished afterwards that I had stayed in the water to see if he would come nearer to investigate; but I was taken by surprise and my only thought was to get back into my clothes and then to the sowar who had my gun with my pony not far off.

When he saw me out of the water on the opposite bank, the panther stood up, looked at me for a moment or two, and then walked leisurely away. Later we tried to track him up but without success. It was extraordinary his being there, three or four miles from the nearest cover, so late in the morning, for it must have been two hours after sunrise. Looking back, what seems so ridiculous is that I felt extremely uncomfortable at being seen in a state of nature—by a panther! I can't explain why: if one had to come to an unduly close acquaintance with a panther's claws, one might almost as well be without any clothes for all the protection they would afford.

I do not think I can bring my stories of the panther to a better conclusion than by giving an account of another experience of a panther in a tree—an incident which for me shed a new light upon the panther's intelligence.

It was among the hills of Northern Gujarat. I had two beats one morning in the hot weather for a panther of no great size: on each occasion she contrived either to break back through the beaters or to sneak past one of the stops. Each time she was marked down again–an unusual piece of luck. The third time was on a hillside overgrown in parts with long grass, in parts with bush of varying thickness, and dotted all over with a number of trees. I took post in a tree at the foot of the hill and the beat started from above down towards me.

About five minutes after the commencement, the beaters were still near the top of the hill when, about half way down, some ninety yards away from me, the panther suddenly appeared, leaping up on to the trunk of a tall tree, bare of foliage, which stood out prominently among its lesser neighbours. Leisurely the great cat climbed towards the summit, and there in the highest fork sat down to watch the advancing line of beaters. For four or five minutes she studied the numbers and dispositions of the enemy, then, having learnt all she wanted to know, proceeded to descend again, leisurely as before, and disappeared in the long grass below.

A friend to whom I told the story remarked: 'Of course she climbed the tree in search of a hiding place'. She did nothing of the sort. No panther's instinct would lead it to try and hide in a leafless tree that obviously afforded no cover whatever, and that when there were actually two other trees with abundance of foliage within a few yards. This panther had been twice disturbed by the beaters: now when she heard the sounds of their approach again she selected the most suitable tree for her purpose and went aloft to reconnoitre. Her deliberation was most marked. Here was no working of that mysterious faculty we call instinct, but an action–or rather a series of actions–of definite intelligence, the outcome of some degree of conscious thought. No man who watched this beast's behaviour as I did could possibly believe otherwise.

And so much for the panther as I have known him.

■ ■ ■

Days and Doings with my Bobbery-Pack

Lieut Col R W Burton (Indian Army)

Year of publication: 1939

What a crowd of memories the above heading brings to mind! Days of intense enjoyment and doings of delight. Up long before dawn, and out of the Cantonment before the short Indian twilight had broadened into day, we used to make our way to the appointed meeting-place where the dogs had been sent out overnight, or very early in the morning. Sometimes it was convenient to ride out, sometimes to drive, less seldom to cycle; and though there may have been occasional disinclination to forego several hour's sleep, yet the pleasure of being up early, seeing the dawn break and hearing the many interesting sounds of awakening life which is the reward of early rising, far outweighed any thoughts of slothfulness.

And that early morning meeting of Master and pack! What a pleasure it used to be. Each dog had its own distinctive manner of expressing delight. *Bob*, the poligar, staunch and stolid, quietly wagged his tail; *Pup* showed her pearly teeth in a delighted grin; old *Nelly*, *Bob*'s inseparable companion, in spite of her twelve years of life, pranced about like a puppy; *Paddy*, his name proclaims his native land, who had always to be led by himself or with one of the opposite sex owing to his incurable penchant for a fight, used to have a special caress, for he was his master's shadow. So the memory runs back a number of years bringing to mind dogs big and dogs small, and of various breeds, for membership was open to all those keen of scent and stout of heart. Dogs, alas, are often but shortlived in this country of so many speedy and fatal ailments to man and beast.

Only two years short of five decades is it since the first nucleus of the Bobbery Pack was established: *Stag*, a Persian hound; *Prince* an Australian; *Jupiter*, country-bred greyhound and a faint-hearted tackler who transmitted his failing to *Jingo*, *Sloe*, and others; *Simon*, most courageous of fox-terriers. These were the forerunners of dogs of varied breeds and colours.

Born in 1868, **Richard Burton**, the sixth son of Gen. E.F. Burton of the Madras Staff Corps, was commissioned from Sandhurst in 1889. Posted to the Indian Army in 1890, he was permanently crippled by a riding accident in 1903. He was, thereafter, assigned to the Cantonment Magistrates Department. A fearless sportsman and a keen fisherman, he wrote over 200 articles on various aspects of Natural History and was the first Naturalist to campaign for the preservation of Indian Wildlife. Col. Burton passed away at his residence at Surrey, England in January 1963 at the age of 95.

Stag and *Prince* it was who, one memorable morning, pressed a buck Chinkara hard for some two miles, running in a circle, while *Jupiter* ran cunning, made a fine effort at a well-judged moment, seized the gazelle at the loins and broke its back with a single bite. Well he knew it would not bite back! It is astonishing how very easily gazelle and antelope are killed by dogs, and yet how tenacious of life they are when struck by bullets. This feat was never repeated by any of my own dogs with an unwounded chinkara as the victim; but old *Nelly*, when at her prime, alone and unaided killed a healthy and full-grown doe antelope among some rocky ground. It appeared to become confused so *Nelly* cut it off after a few minutes coursing, seized it by the neck and killed it instantaneously. She was half Persian, half Afghan, and a very handsome brindle. Poligar *Bob*, before he came into my possession, earned undying fame by seizing a hyaena by the ear and holding it until other dogs came up; how the beast was dispatched history does not relate, probably with a hunting knife. A knife with stout blade of some six inches in length should always be carried.

Beginning in the days of *Stag* and *Jupiter* – *Prince* had to be destroyed as he made the killing of sheep and goats his principal occupation–, many a fine run after fox and jack is seen from the diary to have been had in the vicinity of the Cantonments at which the writer was stationed in those days when the life of the soldier-officer was less strenuous than it is now. Nothing to do after ten o'clock on five days of the week, and Thursdays and Sundays free. Jackals and foxes were numerous; and though the country was much cut up by nullahs, yet falls were few, the hardy and clever country-bred horses and ponies being nimble and seldom coming to grief.

On most mornings a run of some kind was obtained, the tale of *tails* running well into the thirties for the period March to September, when much of the hunting was done. During the cold weather the counter attraction of shooting mostly gave the hunted animals a well-earned respite; a large number of them having been obliged, on one day or another to run for their lives.

At times there were some amusing incidents. On one occasion a spaniel took up a line and ran it hard for a hundred yards, whimpering excitedly as he went. Suddenly he found himself face to face with a big dog jackal. But *Sammy* was a nervy little fellow and quite equal to the tactical situation; he lifted his leg against a bush, pretending that hunting hard on the tails of stray jackals was the last thing in the world of which he was thinking. Seeing me galloping in pursuit, the jack once more set off with valiant *Sam* hard on his heels. A blast on the whistle called up the scattered pack and after an excellent run the jack was duly accounted for.

The dogs very quickly learned to look for signs from Master that quarry had been sighted. All would turn at once towards the sound of the whistle and converge on the galloping pony. When the ground was open I used to have the faster dogs in a double slip-leash so as to give the fleeing animal a good start; for hunting with too speedy hounds is poor sport. Three or four hundred yards is not too much, in open country, to ensure a good run and equalize the chances of hunter and hunted. Foxes can usually take care of themselves pretty well though, owing to their habit of crouching and trying by this means to escape notice, they are sometimes *chopped*.

Some foxes, however, are quite *professional* runners among their kind and have on occasion gone clean away from fast hounds. It is really a very gallant sight to view a little *lomri* skate away with brush held straight behind him after having

My Bobbery-Pack – Bolarum, 1899

extended fast greyhounds to their best efforts for some two miles. This is due without doubt to foxes having a second wind; once they have managed to evade capture until they have gained this second wind they stand a very good chance of escape. In such cases the hounds drop from a canter to a walk and then, as the fox rapidly increases his lead, walk round and round in a state of utter exhaustion with heaving sides and far-protruded dripping tongues. Such defeats do not, however, affect their keenness on future occasions. It is but fair to the hounds to explain that the fox only achieves such a victory after having tired them by many twistings and turnings and agile narrow escapes, even to the skinning of the tail through the gaping jaws of the straining pursuer! It is not uncommon to see a fox jump over a dog's back when hard pressed and necessity forces him to such hazardous efforts at freedom.

One fox there was which raced away on near a dozen occasions before he finally succumbed to *Nelly* after minutes' hard gallop. He must have run twenty miles, one way and another, before he was killed.

Hares afforded good sport, dogs being always keen and excited in their pursuit. They run straight and are very speedy. It used to be a change for the men to turn out forty or fifty as beaters and slip the dogs at hares only. Early in the rainy season as many as eight would be killed in a morning in this manner. Such outings were much enjoyed by the men as well as being good training for them; for they had to control excitement, retain proper intervals, and keep in touch in scrub jungle.

The Indian hare readily goes to ground, it being common for them to take refuge down one of the tunnels of a white-ant mound. As cobras are very partial to these ready made and safe retreats I have often wondered whether the hare may not have been worse frightened when inside than when hunted therein by ravening hounds. His expression would be something like that of the baby cotton-tail *Raggylug* in Ernest Seton Thompson's charming book on animal life.

Escapes from snake-bite were common, but it was rare to have a dog bitten, for instinct seems to warn them not to tackle. A fox terrier belonging to a friend was bitten in long grass and died twenty minutes after the yelp which announced the onset of the tragedy; while another faithful friend, *John Pigeon*, died twenty hours after being bitten while hunting among some rocks. The former snake was doubtless a cobra, the latter probably a Russell's Viper or Daboia. Snakes were seen almost every day, cobras and daboias being numerous. Whenever there was a chance of getting at these deadly reptiles I used to dismount and kill them. Sometimes my horses' legs used to be struck at, but never with success on the part of the snake. Daboias are sluggish but cobras very active and quick. On one occasion a half-bred pointer had a narrow escape. He was trotting along in front of me on a tank bund and suddenly leapt into the air over a snake. The big cobra struck at him, but the agile dog saved himself by an extraordinary effort–a wriggling jump while in the air, which just took him clear of the wicked expanded hood. Having failed in his strike the snake quickly disappeared in the grass.

To return to the hares. One of these once ran 1,200 yards in a perfectly straight line and was killed by the speedy *Dinah* just as it reached cover. The soil was hard and sandy, and the run so remarkably straight that we back-paced it, noting that the hare had not deviated more than a few feet from the straight line during the whole distance.

My Bobbery Pack—Raichur, 1895

Dinah was bought for fifteen rupees at the Arab stables in Bombay, so also *Jack*, a Persian hound, costing twenty rupees. At that time too, *Ginger*, fiery Arab stallion, was acquired for six hundred rupees. A very gallant horse he was; a born steeplechaser, afraid of nothing, and did fourteen miles to the hour in a dog-cart. On one occasion he took off too far away from a mud wall built in the dry bed of a tank, hitting it with his chest. Naturally we parted, but he finished first

Dinah was a true Arab of the desert. Her hereditary ability to withstand thirst was extraordinary; and even in severe hot weather, when all the other dogs would be greedily slaking their thirst, she never drank much, and often did not even do more than wet the tip of her tongue; also her tongue never protruded to the extent of the other dogs. 'Jack' had to be destroyed–rabies–, and *Dinah* was never the same after she produced a litter to *Poligar Bob*.

Only one of the six pups lived and I have never seen *Pup's* equal. She could run rings round any of the other hounds I had. Pace, stamina, keenness, a good tackler, and the very best and most good natured of bitches. In those days I had no camera so have no photograph of her.

Thinking of *Bob* brings to mind that clown Poligar *Jacky* purchased for five rupees from a Horse Artillery syce: he turned out to be a great character. His first exploit was on his entry into the pack when only nine or ten months old. A fox was put up and jinked a great deal, at last running round a rock into *Jacky's* gaping jaws which closed on him, the canine teeth being fixed into the back of the fox's head. *Jacky* was nearly choked by the fox's nose sticking in his gullet so Master pulled his jaws open while *Dinah* removed the fox After this *Jacky* was mighty keen and always a great tackler. Another of his exploits was the slaying of a village porker for which he had to receive a drubbing and I to pay his original purchase price.

THEODORE BASKARAN

A Poligar hound

I parted with *Jacky* and the rest of my pack when going on furlough; a couple of years later I met him in the road and he growled ferociously at my friendly advances. A curious thing about this dog was that he had the smell peculiar to hyaenas; nothing would conceal it; no amount of washing was of any use; so he was banished to the stables. He was a clown of a dog, always playing pranks and amusing even to look at.

There are no better dogs for the Bobbery-Pack than the Poligar. Good tacklers, sound feet, not too fast, stand the heat, splendid staying power, and the best of constitutions. They come from the Poligar country in the south of India. In the early part of the last century the troops of the East India Company had many tough engagements during the Poligar War in the Tinnevelly District, as related by Colonel James Welsh in his Military Reminiscences. At the assault on one place only forty-six of the storming party of one hundred and twenty Europeans escaped unhurt. On that day four officers and forty-nine men were killed. Of one old Poligar it is related that, mortally wounded, he desired to be carried before the British Commander:- 'The old man, who was placed upright in a chair, then said, with a firm voice, "I have come to show the English how a poligar can die." He twisted his whiskers with both hands as he spoke, and in that attitude expired.' So the Poligars of that day were as stout-hearted as their almost hairless hounds.

A cross-bred bull-terrier should be in every Bobbery-Pack. I say cross-bred because pure bred ones cannot stand the heat, and soon knock up on a hot morning. Dog *Tiger* was the best of this kind that I ever possessed. Sixty pounds in weight, with a head like a mastiff, he feared nothing and could pull down anything unaided. He was incorrigible in the matter of pulling down buffaloes. No buffalo, once *Tiger* had him by the nose, remained on his legs for more than forty yards, and no amount of correction ever wholly cured him of this propensity; it was fortunate that he did not molest sheep or goats, for then he could never been taken out. He died of jaundice, that disease so common and fatal to dogs in India.

None of my dogs were really savage. Rampur hounds have that reputation, and perhaps those of a Bihar Rajah were of that breed, though Doctor Daniel Johnson in *Indian Field Sports* calls them Persians, 'the Rajah slipped his Persian hounds after a jackal but when he hallooed on the dogs they mistook the object meant for them and attacked the Rajah's horse, obliging him to ride into the neighbouring river to escape their attack, to the great amusement of the gentlemen present and the Rajah's mortification'.

Dogs of the bull-terrier class are frequently ferocious towards animals though mild and good tempered to human beings; they require to be trained to habits of implicit obedience from their earliest puppyhood or they become a nuisance to their owner and a terror to the community. It is on record that so long ago as the year 1670, a bulldog owned by the Chief of a Factory on the West Coast of India slew a sacred cow (bull?) and the mob murdered every European in the place. The Monument to John Best and seventeen Englishmen thus slain records that: 'They were sacrificed to the fury of a mad priesthood and an infuriated mob.' (*Things Indian* by William Crooke).

Banjara dogs are very good but those of pure breed seldom to be procured, as their owners, that most interesting nomad tribe of gipsy appearance and habits, will not readily part with them. These dogs hunt by both sight and scent and being fast and courageous are suitable for every description of hunting and coursing. I owned half-bred ones now and again.

The best, and also the most pleasant time of year for the sport described is during the early part of the rains before the grass gets high and cover too thick. The fresh feel of the air and a tinge of green all over the countryside is a most extraordinary relief after the heat of the previous three or four months. Butterflies are to be seen daily in increasing numbers; where formerly was parched soil all sorts of wild flowers magically appear; birds are busy feasting on the many insects swarming in the air; crows and kites are greedily chasing and devouring the lovely ground mites with scarlet plush bodies seen crawling over the ground in all directions, and if these are as good to eat as they are beautiful to look at they must indeed be delicious morsels. Then also myriads of flying white-ants issue from the termite mounds, providing a ready feast for foxes and jackals which I have found to have stomachs distended with the easily gorged meal.

It was my custom to always hold an autopsy of stomach contents and at that season jackals were found to be in the habit of eating various fruits, besides the many lizards, rats, mice, and small birds which were more easily come by at that time of the year.

Horses and dogs feel the change as much as their Master, and throwing off the lassitude which has held them during the hot weather months, show by their friskings and gambollings how pleased they are that the season of green grass and cool breezes has once more returned. An animal is sighted in the distance, and discovered after careful scrutiny to be a jackal sneaking homewards to the scrub jungle two miles to the north; so quietly trotting along we lessen the distance to about four hundred yards and then start galloping hard. To the sound of the thudding hoofs and the loud blast on the whistle the dogs come in from all directions, for they soon acquire the habit of scouting to the flanks on their own account, converging to a point ahead of the galloping horse, as they know their Master is invariably to be trusted. Soon the fastest among them gets a sight of the jack; this causes her to exert herself to the utmost and she rapidly gains on the stout varmint before her. He knows every inch of the country and heads for some nullahs and broken ground where he hopes to throw off the panting crew now hard on his heels. Greyhounds, hunting entirely by sight, are soon at fault when the quarry disappears from view and it is at such times that the experience and hard riding of their Master is so invaluable to them.

It is only by hard and judicious riding that one is able, in rough and broken country, to keep close enough to the quarry to enable the dogs to be put on to the line it has taken; so the success of the pack is often largely dependent on the ability of the Master. In the case now in mind the dogs are soon put on and another gallant jack is gathered to his fathers, his mask and brush being consigned to the bag slung at the saddle for the purpose.

Sometimes the fast dogs would be left at home, and the hunt conducted with the slower members of the pack, many of which had quite good noses. That also afforded good sport, and much that was interesting in the way of seeing the dogs at work.

Most of the hunting was around the Cantonments of Aurangabad, Raichur and Bolarum. Wolves were occasionally seen near the two former and hyaenas near the latter. *Tiger* came into my possession at Bolarum so never had a chance at a wolf; nor did he come to grips with a hyaena as it was very seldom one was found away from unrideable ground; so neither of these beasts were accounted for by my pack. One day at Aurangabad, of course when not hunting–I came across two wolves returning, fully gorged to the hilly country from some antelope feast. My camp and servants, dogs, hog-spear, etc had gone on before me so my sole weapon was the cleaning rod of my gun which had been left behind at my bungalow. I rode the wolves to a standstill, actually touching the male with the cleaning rod at which he showed a fine row of teeth. I left them in a pool of water in a sandy nullah and having found my camp returned with dogs and spear, but the animals had left and could not be found. That was an unique opportunity which never recurred.

On one occasion, again no pack and but a rook rifle as weapon, I had a great gallop after a hyaena. My friend on a big Australian pressed the hyaena hard and my much less speedy grey country-bred toiled in the rear. My nag eventually gave out while my friend was unable to use his rifle because of his pulling horse; so the beast made good his escape after a run of some three miles over the broken country east of the Trimulgherry rocks. That gallop lives in the memory.

It is marvellous what reckless riding the fast pursuit of animals induces. I can call to mind one memorable morning when a friend and myself galloped after a jack, in the endeavour to put the dogs on to him, over the top of a rocky hill, across several stone quarries and down a steep, jungly slope on the other side. The jack was a fast one so we failed to get the hounds to see him and had to abandon the chase. A lady was out with us, a nervous rider, so we quietly made our way towards the place from which we had started a mile or so back, being positively astonished at the places we had galloped over and could not attempt to cross them at a walk. At one place my pony had attempted to stop at a perpendicular drop in the sheet rock, but being unable to do so had been obliged to drop some four or five feet on to a slippery surface down which he slithered for some yards before regaining his equilibrium, when he was again urged on his mad career. *Ginger* was indeed a gallant beast and a great handful on such occasions. Truly fortune favours the reckless rider on most days, but the pitcher, not infrequently, goes too often to the well!

I have said little about the horses and ponies. Every man has his own ideas, which must be guided by his weight and inches. For those riding about 11 st. 7 lbs. and under, nothing can be better than the Arab or good class country-bred. Moderately heavy weights can hunt on exceptionally large country-breds but must usually ride Australian animals. Big horses are rather unwieldy for the kind of hunting, or coursing, of which I have endeavoured to give a faithful picture. Handy, speedy, sure-footed animals are best suited to the business, and few would care to risk valuable mounts at such harum-scarum work.

To write of my Bobbery-Pack has been a pleasure, but the narrative must be brought to an end for fear it may become wearisome to readers who are possibly not such enthusiasts as myself. My hunting days are over and I may never again enjoy 'good hunting' except in the memories of the past; but can console myself with the philosophic conviction that once the inevitable has been submitted to there is much pleasure to be had in reflections on days of enjoyment in by-gone years.

■ ■ ■

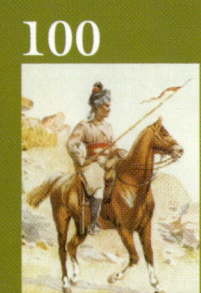

On the Hornbills of India and Burmah

By Lieut Col S R Tickell (Indian Army)
Year of publication: 1864

Lieut Col S R Tickell was an Officer of the 1st Native Infantry and was assigned to the civil administration on the southwest frontier of Bengal. He was later transferred to Tenaserrim in Burma (Myanmar). Tickell was one of the best field naturalists India has known and the manuscript of his Book on Indian Mammals is in the library of the Zoological Society of London. Tickell was also an excellent ornithologist and two birds, a Flycatcher and a Hornbill, are named after him.

1. Tickell's Hornbill *Ptilolaemus austeni* Whitethroated Brown Hornbill

Dimensions, ♀.—Length 2' 2¾"; spread 3' 1"; wing 1'; tail 11¼", exceeding wing by 7½"; bill 4-$^9/_{16}$"; tarsus 1-$^{13}/_{16}$"; middle toe 1½"; greatest vertical depth of bill and casque 2½".

Form.—As in the smaller Hornbills. Neck a little more plumose. A well-developed recumbent blunt crest. Bill and its elevated ridge as in the young of *Buceros bicornis*. Edges serrated by erosion, but without hiatus. Culmen for basal half of bill compressed into a keel-like process, rising rather abruptly from forehead, and inclining downwards and forwards subparallel to the arch of the bill, with which it amalgamates at about 2" from tip, the conjunction becoming more and more abrupt by age, but not exceeding an angle of 45° with the arch of the bill. Orbital space bare. Tail pretty long and rounded, centre exceeding outermost rectrices by 1¾". For the rest the details are typical, as in *O. birostris*.

Colour, ♂ & ♀.—Iris grey; brown next pupil. Bill dark horn; basal half of casque or culminar protuberance dull orange. Orbital skin pale smalt. Legs dark-greenish horn, with pale soles. Head bistre-brown, shafted pale. Upper parts umber-brown, rather dull and opaque, with a tinge of olive, and glances of dull green in half-lights. Secondaries and primaries greenish black, the latter with their outer margins midway, for a short space, and their tips whitish. Two central rectrices as back, with pale tips; rest greenish black, with pale tips. All under-parts ferruginous, rather pale, brightest on throat, dull and clouded with vinous ashy on belly.

The district of Amherst, in the Tenasserim Provinces, is longitudinally traversed for its whole extent, north and south, by a range of mountains (a branch of the Yomatoung of Burmah), which proceeds southwardly, through Tavoy and Mergui, into the Malayan peninsula, of which it forms as it were the backbone. The range is composed of numerous ridges, more or less tortuous, rising along the middle to peaks of from 6,000 to 8,000 feet in height, and occupying in breadth a space of about 40 miles. The hills are exceedingly steep, with narrow profound valleys, and everywhere clothed with dense forest and underwood, except on some of the loftiest summits, which are bare granite, scantily clad

with coarse grass and scrubby bushes. But on the lower spurs, and especially in the plains at their feet, the soil, watered by numerous brooks and streams, and covered by vegetable detritus washed down from the uplands, is exceedingly rich, and nourishes the growth of trees, which attain prodigious dimensions. The "Thengan" (*Hopea odorata*), "Toungbing", and "Kathykkha" trees, used by the Burmese and Taláings for making "dugouts" of 50 tons and upwards, rise to 150 feet before sending out a branch, their summits attaining a height of 230 feet, and their trunks smooth, round, and perpendicular, measuring near the ground from 10 to 12 feet in diameter.

It is exclusively on these giants of the forest that this species of Hornbill rests and feeds, never being met with in jungles where the trees are of ordinary size. I met with them from about the base of the hills to as high up as 4,000 feet above the sea-level, but not beyond. They appeared less rare on the easterly skirt or Siam side of the range, occurring in pairs or small parties of five or six, incessantly calling to each other in loud plaintive screams, "*whey-wheyo, whey-wheyo*", and while engaged in feeding, keeping up a low murmuring cackle like Parrots. Their flight is smooth and deliberate, and it is performed at great elevations, especially when crossing over from mountain-top to mountain-top. Keeping thus ever at immense heights, and being withal as quick-sighted and wary as the rest of the genus, it may readily be imagined how difficult this bird is to procure with the gun. I succeeded, in fact, in "bagging" but one specimen, and wounding another, which escaped, during my cold-weather excursion into the Tenasserium Mountains in January 1855. I procured two more, some years subsequently. In the case of the specimen, its companions showed much excitement when it fell, coming boldly down to the lower branches, with loud screams, and remaining within easy shot while I was reloading. This occurred at Thengangyee sakan (literally, "halting-place of great Thengans"), a spot in the forest so named from the huge Thengan trees about it, situated on the eastern skirt of the range above described. This is one of the resting-places on the wild path pursued by travellers from the Shan states of Yahan, in Siam, to Moulmein. On revisiting the same spot in March 1859, not a single bird of this kind was to be found there, or in the hills around. Being a frugivorous bird, it has to make partial migrations, as its food fails or passes out of season in one place, to where some other kind of fruit, is ripening—a compulsory habit, common also to all the fruit-eating Pigeons. I fell in with them, accordingly, during the last-mentioned period, in a very different locality, in the flat forest lying along the south of the Houngthrau River, considerably to the south of Thengangyee sakán, and on a much lower level. They were on these occasions so wild as not to allow approach within gunshot; but on my last day's march, which led through these forests to the banks of the Houngthrau, where my boats were in readiness to take me down to Moulmein, I came across three of these birds near a Karén clearing. To my surprise, they allowed me to approach within a long shot. The first bird I fired at fell from his perch (on a vast tree) into a thicket. Its companions did not fly away; and my second barrel brought down another, which hitched in the tree, to all appearance desperately wounded. I, of course, congratulated myself on having secured two of this very rare species; but, to my intense chagrin, when my people had come up to search for and secure the prizes, the second bird flew away as if unscathed, and the first was not to be found! The Karéns declared they had seen it fly away! And, in truth, the voices of all three were presently afterwards heard from the interior of the jungle. The heavy morning dew made any diversion from the path equivalent to a plunge in the river; so I sent a Burman follower, whom I had taught the use of a gun, after the fugitives, and he succeeded in fairly bagging two of them. They proved to be males, not differing perceptibly in plumage from the female.

Tickells   Hornbill
Whitethroated Brown Hornbill
Ptilolaemus austeni

Great Pied Hornbill *Buceros bicornis*

2. Great Pied Hornbill *Buceros bicornis*

Of this Hornbill Hodgson has left little to say. I have kept several specimens alive, and have been an eye-witness of the singular mode of incubation of the bird. The young have the casque no more developed than in the subgenus *Toccus* of Lesson. At the commencement of the second year the anterior extremity begins to separate from the culmen, and during the third year assumes the transverse crescent shape, sending the two edges or cornua outwards and upwards, while the whole anterior portion gets broader, till it is equal to the hinder part. But the casque is not fully developed till the fifth year. Nevertheless the brittle and quasi-osseous edging to the bill is perfected in the second year, becoming quickly eroded by wear and tear. In Nepal, according to the natives, the "Homrai" or "Ban-rao" (King of the Woods) ascends the mountains to near the snows during the hot weather. In the Tenasserim Provinces, however, I have found it so late as April in the lowest and hottest forests, and never higher than 2,000 feet above sea-level. Our *hill* Hornbill in that country is *B.* (*Aceros*) *nipalensis* of Hodgson, which I have shot on a spur of the great Mooleyit peak, full 3,500 feet above the sea.

The voice of *B. bicornis* is prodigiously loud. Its roars re-echo through the hills, and it is difficult at first to assign such sounds to a bird. As in other species of which the notes are sharper, the noise is produced both in exhaling and inhaling. *B. bicornis* when caught

RAJAT BHARGAVA

young is easily tamed, but becomes bold rather than gentle, menacing a too near approach with its huge bill, which inflicts severe bites. Those I have had in my possession would not suffer themselves to be caressed. They flew about the garden and grounds, resting on large trees or the roof of the house, and often coming to the ground, where they progressed by sidelong hops, squatting occasionally on their heels (or elbows) and searching for food in the grass, where they picked up and swallowed insects and worms. I once saw one of them seize a frog; but after nipping it and tossing it about, the bird relinquished it. Early of a morning, when the dew was heavy on the ground, I have seen this bird go flapping through beds of weeds or long grass till thoroughly saturated, when it would sit in the sun, with expanded wings, drying itself like a Vulture or Cormorant. This species has a singular palsied jerk of the neck in moving the head from side to side or up and down—a peculiarity owing perhaps to the rigidity of the connecting ligaments of the cervical vertebrae, as described by the late Dr. Bramley when residency-surgeon in Nepal. In a captive state, I have never heard this bird utter more than a little murmuring grunt. Its capacity of swallowing is prodigious: a whole plantain can be gulped down without an effort. In picking fruit off a tree, it tosses it up into the air, and lets it fall down the throat. It eats lizards readily not only from the hand, but will search for and seize them. The unfledged or half-fledged nestling constantly utters a feeble croak, alternating with a piping, whistling noise. A remarkable trait I observed in one or two of the birds in my possession was their fondness for rain. They would remain for hours exposed to the heaviest shower, and sit perfectly saturated, with the water trickling from the end of the beak in a ridiculous manner.

Oriental Pied Hornbill
Anthracoceros albirostris

3. **Pied Hornbills**, *Anthracoceros coronatus, Anthracoceros albirostris*

These two nearly allied species inhabit the forest respectively of India and of British Burmah. The second-named is the more numerous of the two, as far as my own observations lead me to judge. I have met with *A. coronatus* in the forests of the jungle mahals, Midnapore, Singbhoom, and Chota Nagpoor; but it is by no means common. *B. albirostris*, on the eastern side of the Bay of Bengal, is much more frequent. The manners of the two species are so much alike, that they hardly need separate description; with *B. albirostris*, however, I am now familiarly acquainted, having had two or three of them tame, in confinement and at large, for this bird becomes so soon domesticated as not to require imprisonment, if it be brought to the house from the nest. It remains perched on a verandah-rail, soon becoming accustomed to the inmates, and readily takes food from the hand. One or two pet ones are to be seen in almost every village in Arakan. Those I possessed used to roost on the roof; flying in and out of the house at pleasure. One in particular, which we kept for nearly two years, became a great favourite. It was fond of being patted and stroked, and would beg for the luxury, throwing its head back to have the throat scratched or tickled. During the day it usually sat under the portico, hailing every arrival with loud screams, and unalarmed by the noisy approach of carriages. It would fly to me or to the children from any distance in the garden or grounds; and especially attached itself to the young folks, allowing them to scratch its neck, throwing itself into absurd attitudes, as if coaxing them to continue, and never on any occasion hurting them with its formidable bill. Thus the bird, grotesque and ugly as it was, became associated, in our minds, with its pretty playmates, and an inseparable appendage to their little sports. But alas! Like all pets, this one met with an untimely end. It was found by the servants early one morning on the roof of the house, dead, with the marks of teeth in its breast, inflicted probably by some marauding cat which had surprised it during sleep.

Malabar Pied Hornbill
Anthracoceros coronatus

Like the rest of the genus, *A. coronatus* is nearly omnivorous, but prefers fruit to other food. According to Jerdon, it is found in Northern India, Midnapore and Rajmahal, and Monghir on the Ganges. My own experience does not corroborate this. The breeding-time in Arakan and the Tenasserim Provinces is in July or August; and the female is said to lay two to four eggs in the hollow of a tree, without any nest. The eggs are white. I have never seen them.

4. Wreathed Hornbill *Aceros undulatus*

This species is very numerous in the Tenasserim Provinces and in the inland forests of Arakan. Its presence is soon known, on entering the lofty woods to which it resorts, by its loud and as it were menacing voice, uttered in a short, gruff dissyllabic croak, "*Kukkuk*", which it repeats at intervals, either when perched or when flying over the tree-tops. Its powers of flight are much more extensive than in any other species of the genus. I have seen parties of five or six of these birds in Arakan, high in air, flying over the sea; and have watched them till they melted from sight into the horizon, as if they had finally left the shore. Where these excursions end it would be curious to know; for the bird is not found on the west shores of the Bay of Bengal. Its most northerly habitat appears to be the hilly jungles on the highest parts of the Koladyn River in Arakan; but whether it extends into Chittagong or the hills of Cachar and Manipur I know not. They seem to get more and more numerous towards the south, and on the Houngthrau River, which rises in the southernmost Shan states of Siam, are quite common. The flight of this bird, unlike that of the species before described, is slow and regular; and the rush of the air through its pinions so loud as to be heard at half a mile distance. This remark applies also to *B. bicornis*. It is generally wild and wary; at times, however, when feeding on the fruit of some large *Ficus* (its favourite resort), it will allow approach within gun-shot, and is so voracious as to return two or three times to the same tree after being as often shot at. They settle generally on the large branches near the summit, jumping from place to place, and greedily picking off and swallowing the soft ripe figs of the Banian, Ber, Goolur, Peepul, Pákhur, or similar trees of the *Ficus* family. They feed usually in silence, and mix indiscriminately with the numerous fruit-eating Pigeons and Monkeys which, similarly engaged, constitute a singularly varied crowd amongst the lofty branches of these gigantic trees, and a picture so remarkable as not to be easily forgotten by the observer.

The female incubates generally about the end of the cold weather, laying two or three eggs in some convenient hole high up the stem of a tall tree in the deepest forests. The Karéns say that the female is not immured while sitting, as is the case with *B. bicornis*. An egg, brought to me towards the end of February 1855, is pure white, opaque and coarse on surface; size 2" by 1½". I was on that occasion on my way down the Houngthrau, a clear, pretty stream, shaded by lofty timber, eddying in deep pools under high gravelly banks, breaking into foam and tumbling over boulders of sandstone, or rippling along shallow beds of clean pebbles and silvery sand. To the last-named spots, just before or during the short twilight of a tropical evening, these Hornbills used to resort in great numbers, allowing my boat to approach pretty near, as it glided down the stream. I could thus watch them on the little sand-flats, hopping freely enough along the ground, and delving their beaks in as if searching for worms or mollusks; while some stood up to their bellies in the water, apparently much enjoying their bath. As the dusk gathered over the river, I remarked them resorting to roost on the loftiest trees fringing its course. The Karéns who live in these virgin forests say that between the "Yowng-Yowng" (*B. bicornis*) and the "Owkhyen net" (the present subject) there is always open war; and, in truth, I do not remember to have remarked the two species anywhere together.

Wreathed Hornbill *Rhyticeros undulatus*

A Mahseer River of Southern India

Lieut Col R W Burton (Indian Army)

Year of publication: 1939

The majority of visitors to the 'Blue Mountains' of Southern India see the Bhavani river at Mettupalayam, where it is spanned by road and rail and is a very ordinary looking stream of no particular attraction; yet, but a few miles up the valley, it is a rapid and beautiful river of many moods.

Those who come to the Hills from the direction of Mysore cross the largest tributary—the Moyar—which is, during some twenty miles of its course, one of the most inaccessible rivers in India, for it runs at the bottom of a tremendous ravine close upon a thousand feet deep known as 'The Mysore Ditch'. Apart from this natural obstacle the river there passes through forest country inhabited only by wild beasts and jungle tribes, and the steamy depths of the gorge are protected by a malaria said to be deadly even to the aboriginal people, who dare not remain on the river banks after sundown.

There are monster mahseer in those wonderful pools and deep runs, as has been discovered by a few adventurous sportsmen; but malaria—the most efficient and ever-watchful of all Game Wardens—preserves the wild beasts and the fish from the all-destroying hand of Man. To some extent the Bhavani is similarly protected for the first ten miles above Mettupalayam, but for twenty-five miles above that it is a lovely fishing river.

An affluent of the mighty Cauvery, the Bhavani joins that river at the town of Bhavani (another name for the Hindu Goddess Parvati, also known as Durga—The Earth Mother) which is twenty-four miles below the recently constructed Mettur Dam. This junction is some sixty miles below Mettupalayam and forty miles after the Moyar comes in on the left bank.

Through the writings of H.S. Thomas, whose work *The Rod in India* is as deserving of immortal fame as Izaac Walton's *Compleat Angler*, so charmingly is it written and so full of wise advice, this river is known to many by name; and the terrible reputation the lower gorge of the valley had sixty years ago for malaria obtains even to the present day. The fact is that along some twenty miles of the wider portion of the valley there is little malaria during the months of August and September, though all the usual protective

Born in 1868, **Richard Burton**, the sixth son of Gen. E.F. Burton of the Madras Staff Corps, was commissioned from Sandhurst in 1889. Posted to the Indian Army in 1890, he was permanently crippled by a riding accident in 1903. He was, thereafter, assigned to the Cantonment Magistrates Department. A fearless sportsman and a keen fisherman, he wrote over 200 articles on various aspects of Natural History and was the first Naturalist to campaign for the preservation of Indian Wildlife. Col. Burton passed away at his residence at Surrey, England in January 1963 at the age of 95.

precautions have to be taken for oneself and the camp followers, while the myriad mosquitoes and other biting insects which greatly mar many a fishing trip are entirely absent.

Is there then no snag—apart from the many in the river—to this Angler's Paradise? There is—the necessity of enormous patience! At other times of the year the water is either too low or always in flood, but during these months mahseer are in all the pools and rapids. The fisherman must, however, have both time at his disposal and the patience to wait, as there may be many spates during which no fish can be caught and the angler has to assume the role of *rusticus expectans*. On the occasion of his first visit the writer sat seven days in his tent *dum defluit amnis*, for heaven's flood-gates were open all the time. It is only one who is 'passing his pension days', as the Aryan brother expresses it, who can spare the time for such idle occupation.

Let us get to our furthest camp which is a mile beyond the Irula village of Seerakadavu. Here the river leaves the mountainous country which it has traversed from its sources on the southern slopes of the Kolaribetta Hills (8,624 ft. and only 26 ft. lower than Dodabetta, the highest point in the Nilgiris), and gathering to itself many streams well known to trout fishermen, descends through dense and malarious forests and a precipitous gorge to the village of Attapadi (The Abode of Leeches)–where it makes a left hand turn to shape a very first direct north-easterly course for Mettupalayam. So it can be well realized why the South-West Monsoon occasions so many floods and freshets to the annoyance of the fisherman, though but little rain falls in the valley itself at that time of the year.

Within the angle above mentioned is the conspicuous hill named Malleswara (5,458 ft.), with its remarkable pinnacle having a 400 ft. perpendicular scarp, which

A.J.T. JOHNSINGH

dominates the whole valley and can be seen stabbing the sky from Mettupalayam. It is said that no man has ever set foot on the top of the pinnacle though it looks not too difficult of access by the eastern slope. On the opposite summit is a shrine visited by the Hindus of the countryside during the month of August. It can be conceded to the perspiring pilgrims that they have indeed earned 'merit' as a reward for this toilsome journey, which no women are permitted to perform.

In the upper reaches above this camp the river narrows greatly, and as the fish mostly run small those waters should not be fished. Ordinarily one returns to the water all fish below about six pounds. A beginning is made with Camp Pool and Camp Run which empties the river into a long, deep tree-bordered stretch of water. In Camp Pool are some bad snags. Several fine fish up to fifteen pounds have been taken from it; and one good one bored to the bottom, below the heavy water spilling into the pool, so boulders were heaved in to dislodge him. Suddenly He's off—(not the exact exclamation!)—and it was found that the steel link next to the bait had snapped. Probably one of the stones was too well aimed for otherwise this breakage, not taking place at the moment of seizure, is difficult to understand. The day before fish of seven and twelve pounds were killed, and the next day a heavy fish took a crocodile spinner in the deep pool and tore off one of the wire-mounted hooks. It is seldom artificial spinners are used for they often fail one and are more liable to catch in snags, whereas one treble hook, mounted on wire and used with a small dead bait as Thomas directs, has the least chance of mishap; and if properly adjusted the spin is more natural and effective.

That same day a fish with an edge to his appetite took the bait almost as it touched the water when popped over a rock; the wire trace was in twirls from the pull and the reel screamed. In August this year (1939) an unusually heavy fish for this part of the river was killed. Taking the bait—a four-inch fish—in shallow rapid water above a long, deep pool, he raced a full hundred yards before the coracle could be started after him; and another fifty yards before we were properly feeling him. Then all went rigid—not a move—and after some time it was realized that the line was hung up in some way. All that could be done was to reel in with rod top touching the water beneath overhanging branches, the coracle held by a man on the bank. Then the boatman dived with hand on line and found some fifteen feet down that it had passed under a projecting snag. The relief of mind when the line sprang to the surface! But we were not yet out of the woods, or the fish out of the water, for he was fifty yards further down and could not be felt; another snag. This time he was moved by rod work, and having been well rested did a hundred yards' sprint to yet another snag. How well he knew all the safe anchorages! There he was soon dislodged, and after further down river cavortings was gaffed after a full hour at least five hundred yards from the start. Nose to tail-fork 43 in., girth 27 in., weight 41 lbs.; a fine cock fish; and the hook dropped from the roof of his mouth as he was laid on the ground!

In the remaining water from this camp, extending for a mile or more as far as Village Pool, fish have been taken on various trips, the best being 18 lbs., and one or two certainly nearing the thirty-pound mark have escaped for one reason or another; it is mostly snag-like arguments which cause these lamentable partings.

Storms of the N.-E. monsoon along this river are very violent, as evidenced by the many charred and riven giant trees along the banks and lying in the pools and rapids. Occasionally there are heavy storms from the South-West. In the late afternoon of one 30th August when at this place—'Top Camp' I call it—the sky became overcast and thunder growled among the hills. The great sky-piercing tooth of Malleswara was lost to view and the hills changed from purple to pitch black. Soon the darkness deepened into night and the fireflies spangled the near-by bushes. Nearer and nearer came the storm,

heralded by loud thunder claps and long flashes of vivid lightning. Now the thunder spoke in a long menacing growl and the incessant lightning disclosed a turmoil of leaping waters, for the rapid mountain torrents had already brought the river down in spate. The big trees lining the river banks swayed before the rushing wind, and in a few minutes the camp was flooded deep with torrential rain.

It had already been arranged to move camp down the valley in the morning so the coolies found the sodden tents rather more than they had bargained for; a small extra payment and gift of chewing tobacco heartened them for the nine mile march to Koorapati, which is a favourite camping place opposite where the Siruvani, joined a mile above by the Gopanari, comes in on the right bank. These rivers are a great nuisance. They drain a tea and coffee estate area so become turbid on the slightest excuse (an instance of the great denudation of soil caused by planting operations), so all the lovely length of the Bhavani from here to the Kundah River junction, ten miles down on the left bank, is often unfishable; and the Kundah carries on the bad work all the way to Mettupalayam

The river being in spate we must switch off to another year. The coolies having been started off, I get into the coracle to journey down the river, fishing all the best places all the way. Fish of ten to twenty pounds are usually found at certain known spots. One favourite place, where the Varagaiar stream runs in on the left bank from its mountainous descent, has now been spoilt for many years by the fall of a forest giant which completely snags the incoming water. I doubt if it will ever be shifted. Leaving the Varagaiar Pool, which ends a two-mile stretch of deep and mostly placid water, containing fish 'as big as bullocks' say the 'locals', the river becomes shallow and rapid for some two furlongs. In this length mahseer up to ten pounds may be expected and an occasional murrel: good eating, poor sport.

While the coracle is being taken down some turbulent water it may be given a few lines of description. It is rather wonderful what it can do. Being but six feet in diameter and fourteen inches deep yet it will safely ferry five men and several hundred pounds of

Colour varieties of Mahseer

baggage. The weight when dry is 45 lbs. and one man can carry it for miles without difficulty. The covering is of buffalo hide stretched over a basin-like basket of bamboo. There is but one seam, and the leather being kept in condition by applications of mustard oil lasts a long time. The resilience of construction admits of any amount of bumping over and against boulders, and any damage is easily repaired.

My fishing seat is a hinged-lid packing case 19 in. by 15 in. by 10 in. high, with padlock. A leather-covered cushion is necessary to soften the rigours of the box, or what naturalists call 'ischial callosities' will result. Fish up to 15 lbs. or so–two of them–go into the box; more, or bigger, stay outside. The whole affair–coracle and seat–costs twenty rupees: if one goes to another river the framework is left behind and a new one quickly made at the next camp. Hire of coracle, boatman, cooly, and boy to cook their rice is less than three rupees a day.

On arrival at Koorapati camp the tents are found pitched and breakfast ready. Let us have a look at our surroundings. The valley is very green just now, and along the length of it are the beautiful Nilgiri Hills down the slopes of which the footpath brought us over 3,000 feet in three miles, from a temperate climate to a more than warm one. The inhabitants of the valley are Irulas, an aboriginal people now busily ploughing their stony fields for sowing the ragi crop. They are small in stature and very dark skinned; their noses are broad, and the people of this tribe are so like the Kurumbars that they can with difficulty be told apart. The men wear a top-knot, and the women wear their upper cloths stretched straight across the breast and passed under the arms. Their ornaments are brass bangles and necklaces of beads, so the similar goods from Woolworth's presented by me from time to time have been met with broad smiles. The numerous children appreciate scrambles for pice.

The Irulas will not eat beef. Each village has its cemetery, the dead being buried in a sitting position with the legs crossed tailor fashion. A lamp, knife, and hatchet are placed beside the body. The huts the people live in are but little different to those they run up in a few hours for the visiting sportsman and his servants. Their physical condition is poor, and it is curious that they should remain in this primitive condition within but a short distance of motor traffic and the civilization of Mettupalayam.

Except along the river banks, which are fringed by large evergreen trees, the jungle of the level country and the lower slopes of the hills is thorny, abounding with cactus and several species of euphorbia. Snakes are numerous. The wife of a river watcher was bitten by a viper a while ago, but happily recovered. It was a small snake, the husband said, so more than likely *Phoorsa* (*Echis carinata*) was the culprit, for there was internal haemorrhage a few hours after the bite and the local symptoms were very severe. The woman was in a great fright as there had been several deaths in the valley from snake-bite.

On the right bank of the river there is Government Forest which contains a few animals and some jungle fowl. One hears no sounds at night so the few tigers must live on cattle. This side of the river the Irulas have killed off everything and now complain of having no meat ! A tigress killed a cow about three hundred yards from my tent at this camp on one occasion, but being an inexperienced young female did not make a neat job of it so the bellowing protestations of the poor cow brought shouting villagers to the scene and the beast went off without her dinner.

Now that the storm has ceased, the sun is drying up the earth almost as fast as it was saturated by the rain. Those exquisitely coloured insects, with soft plush-like oval bodies about the size of a small cherry stone but more flat, are making their appearance. Soon they will lie about so thickly in places that they will redden the earth. Touch one and it will remain quiescent, a method of self-protection common to many things which creep and crawl. Brilliant butterflies are again flying about and one wonders how they escape damage during such boisterous weather; some indeed are bedraggled, and some have bits of wings clipped out by pursuing birds, but many are intact. Dragonflies and other insects hawk the rain-washed air; the song of birds is among the trees; hawks and vultures circle in the sky; all nature is smiling, which means that everything is more than ever busy eating everything else!

The river being again fishable an hour's walk takes me to 'Top Pool' by half-past eight to begin fishing by nine o'clock. Three four-inch fish from the bait-can are carefully mounted and the spin of each tested along the water's edge. Long practice makes perfect, and it is but seldom the spin is not satisfactory.

The bait must spin quickly for, as Izaac Walton wrote some 280 years ago in regard to spinning minnows for trout: '. . . try how it will turn, by drawing it across the water or against the stream, and if it do not turn nimbly, then turn the tail a little to the right or left hand, and try again, till it turn quick; for if not, you are in danger to catch nothing; for know, that it is impossible that it should turn too quick.' I make sure that the spin 'remains put' by means of sundry sewings with strong thread, white or khaki it matters not, and the coracle is quietly paddled to where the Varagaiar rapids come in with a pleasing turmoil.

Away the bait goes with a Hillman bullet of suitable weight to keep it well down, and the cast being across and a little upstream it sinks to somewhere near the floor of the river and swings round with the current. Again and again this is repeated, the coracle quietly drifting, until a cast into the very edge of the current swirling along the steep, tree-rooted bank on the further side results in a furious tug, and that thrilling scream of the reel denoting a heavy fish. The tail of the pool is a full hundred and fifty yards away and he stops some eighty yards short of that, thus leaving that most promising place undisturbed. The first run safely completed the fish is humoured by steady strain to work upstream a bit, and so by degrees, after several other long or short runs and sundry shakings of head, is gradually tired out. Soon we see a surface swirl; there are tugs and short excursions which must not be too strenuously resisted; the coracle is beached against the sandy shore, the gaff made ready, and after some twenty minutes of play a fine hen mahseer of twenty-two pounds is safely laid on tile bank. A lovely fish; mahseer are a never ceasing source of delight to the eye and the mind of the fisherman in the East.

Again we set out, but this is on another (day in another year) and at the tail of the pool where the escaping water divides either side of a jâmun bush the questing bait is taken by a heavy fish. The coracle is committed to the pull of the left channel and the fish prefers the right! Quick work with the rod takes the line sliding over the bending twigs of the first bush but the fish is too fast to permit of similar escape from one a few yards below, so coracle and fish race down the rapid water, the latter an easy winner.

Obviously the coracle must be swung to a halt against the left hand boulders and bushes as soon as possible, which is some eighty yards from the start; while the fish is quietly resting in the deep water of the pool a hundred and twenty yards from where hooked: the line is held taut by the twigs and the fish can have fifty more yards of line if he wants it, so all is well—perhaps! Two men, a fish watcher and the coracle cooly, have a difficult arm-clasping-arm crossing through the strong water of the run and make a diving plunge at the last moment to seize a saving branch. Soon the line is passed over the bending twigs and a furious winding in finds the fish still on; it puts up a stubborn fight in the strong water but is eventually tired out and brought to the sandy bank beneath the giant trees where he soon shows his broad bullock-like back and bright red fins. A deep and heavy fish of thirty pounds. Such happenings as this almost bring on a heart attack! It was noted on that occasion that a multiplying reel must be the next purchase.

After this pool—I call it Lower Pool—there is a most promising run into a long, deep stretch of water—Halcyon Pool—but only once have I taken a fish in it and that only six pounds. Then, after another portage, comes Mug's Pool, so named because there has so often been an easily taken seven-pound fish at that junction of two channels of the river. Then down a heavy rapid to Falls Pool which is formed by a rocky barrier across the river. It is at the tail of this, where it narrows towards the left banks, that one always expects a good fish. These come up Grandfather's Run from the deep pool below to lie on the sandy floor and feed on whatever the strong stream may bring down. When the river is 'just right' it is with confident expectation that the bait is sent to search out every yard of that lie. Fish of twenty to twenty-seven pounds have been hooked there many times, and all have to be followed in the dancing coracle to the pool for to attempt otherwise spells certain disaster. Where the long and turbulent run ends big fish may also be expected.

Now comes a lengthy portage—except when the river is high and a heavy rapid with three feet waves is negotiated—taking one to Temple Pool (a small, primitive jungle shrine in among the lofty trees) and fish not above fourteen pounds; below that a pool where a fish of eighteen pounds was taken on the first trip nine years ago (no other at that place since then at that seemingly excellent lie), and so down runs and rapids and long still pools in all of which a few fish have been caught from time to time, but only parts of it really good water, to Koorapati Camp.

In the intervals of searching the more likely places with the spinning bait there is much of interest to be observed and charm of scenery to delight the eye as the coracle moves down this lovely river. There are long deep pools shaded to the water's edge by evergreen trees within the shadows of which are widening rings and splashes of surface-feeding fish; amidst the foliage birds and monkeys are descried, and sometimes a large maroon and buff squirrel is detected through the dropping of a fruit from the topmost branches of a giant tree; pied hornbills pass over with peculiar undulating flight; kingfishers of three varieties are perched or hovering; a snake-necked darter rises spattering from the water; a fish eagle is perched on the bare out-stretched branch of a tree; a disporting school of otters may be seen, and at certain known places a crocodile is looked for.

Fitful air movements run light wrinkles over the water, while an occasional puff of wind causes a flash of darkness to sweep across the placid surface. Looking upstream the vista of the wide waterway framed by the forest trees is backed by the dark green of the streaming ravines set off by the grassy or jungle-clad slopes of the billowing hills. Sometimes all is clear and detailed to the sight, then a turn of the river affords a distant view where the receding hills are clothed in a tint of the finest purple. The delicate blue surface of the water reflects the azure of the sky; each turn of the river discloses an ever-changing scene and over all is a feeling of perfect peace. Yet all the while one is sensible of that tragic world which lies beneath the surface of the waters, where there is no thought of mercy for any living thing which can neither flight nor flee. Nature, while smiling and beautiful to the human eye, is full of horror based upon the killing of things, sometimes with great cruelty, and seemingly with infinite futility and wantonness, as is instanced by the immense destruction of fishes' eggs and small fry; but nature does not ruthlessly exterminate as does the ever destroying hand of man. Some thirty years ago this river was almost entirely denuded of fish life. Only the provision of fish watchers by the Government has restored it to something approaching what it should contain.

After this long discussion we set off at three thirty—the Siruvani being almost clear—to fish the very excellent downstream runs and rapids which have produced many fine fish. Close below camp is a good place denied to me for the past nine years by many snags; it may perhaps never be fishable, so firmly are the great trunks and branches interlocked. Then comes a long, narrow channel in which I have taken fish from ten pounds to fifty pounds. At the tail-end, where the heavy run curls over to break into boisterous waves is the place where hope runs high. In 1935 eleven fish were taken from that channel, the best being 51 lbs. (45 in. to tail-fork, 29 in. girth). That fish was unfortunate for he was held by only one strand of the three-ply wire trace and the points of the hook had not penetrated beyond the barb. There were many anxious moments during the hour of play.

The spinning rod on that occasion was a 10 ft. 6 in. green-heart; for parts of the river where fish below fifteen pounds are expected a 7 ft. rod is suitable.

A 26 lb Mahsir

Below the channel is a mile of good water where fish are always found (Siruvani permitting) which ends in a deep pool where big fish lie. From here to Sundapati is four miles of pools, runs, rapids, and shallows but no fish over fifteen pounds have been found in it.

The travel fast downstream is a great delight. There is the shallow transparent water where the always beautiful mosaic floor of the river rises to the eye. Every pebble, every tiny fish, can be clearly seen. Now comes the increasing speed of the water sliding into the turmoil of the rapidly descending torrent; the rushing sound of the wave-tossed rapid; the deep growl of the angry waves obstructed by a boulder, and the exhilaration of a slight element of danger as the dancing coracle is dexterously twirled to pass the peril of a surface-breaking rock: then, shooting down at a speed which is faster than the stream itself, descent is made into the long reach of waves and eddies merging into sullen swirls like richly watered silk as the chastened stream mingles with the placid waters of the next long pool.

These entrances and exits of the pools are where the big ones lie; but unless the harbouring pool is really deep no fish of size will be found, for the mahseer likes his rest in the intervals of feeding excursions to be wholly undisturbed by surface happenings.

At Sundapati is some deep and heavy water and a splendid outgoing run some fifty yards wide yet, for some reason unknown, it is seldom one gets fish there. That there are some is certain, but those great-grandfathers do not come to the spinning bait: too lazy perhaps. In 1900 a Colonel MacArthur fished the previously baited pool with ragi paste on his large hook and killed a mahseer of 92 lbs. which was much admired and appreciated by Boer Prisoners of War in camp near Wellington. The late Mr. Mark Clementson of Ootacamund killed fish of between sixty and seventy pounds in the same pool by the same style of fishing, which is the method by which all the hundred-pound mahseer one hears of from Mysore waters of the Cauvery and Cubanny rivers are taken. Heave afar a large lump of ragi paste and sit expectant while the mosquitoes eat you and the fish eats the paste! Good enough when coracle and spinning rod, or fly rod, sport is not possible.

It is after a week or more at Koorapati that one moves down to Sundapati, from where about three miles up-and downstream can be fished. Possibly the sport here would be better in the warmer months of January and February. Certainly there would be good fun with the fly rod, for Carnatic carp run to ten pounds, and there are numerous other fly-takers also.

Sometimes, when the weather during August-September has been warmer than usual. I have had good sport with the fly rod in the long, still pools. The coracle is quietly floated down the centre of the stream and casts made under the shadowing trees and close to the steep banks. When the flying white ants emerge in their countless thousands many of them fall into the river to be gulped down by the ravenous fish, for the water

will boil with rises when this luscious feast is served. A fly to imitate this insect has recently been put on the market with the name of 'Gibby's Ant', and is excellent for use at that time. For ordinary times there is a variety of flies—mostly dark ones; and a small green caterpillar on a worm tackle, the hooks considerably stronger than those used for trout, is very effective. The rod should be about ten to eleven feet in length, and casts of suitable strength.

Six miles below Sundapati is the Kundah river junction, and for almost a mile before that there is good water holding big fish. The Kundah has its sources in the high hills and comes in by many shallow channels, while below is a deep, rocky run bare of trees and vegetation where there are at times big mahseer difficult to entice to a spinning bait. I have seldom fished there.

Now the river descends steeply in the series of roaring rapids, cascades and waterfalls, quite impossible for the coracle, into the deep malarious gorge mentioned early in this article. I do not know if anyone has really fished there. It would be very exhausting and difficult work in those steamy depths and safe foot gear would be very necessary. I have just come across the perfect boot for slippery rocks, whether these be under or out of the water. They are made of rubber uppers and soles of compressed felt. Perhaps I may some day essay that gorge.

At the foot of the gorge is fishable water which can be got at from Tekampati situated two miles by bridle path from the river bank. *Rod in India* should be read as to this part of the river; 'fish as big as portmanteaus', wrote Thomas. Arrangements from Mettupalayam, six miles distant, would present no difficulty. There is a motorable track.

Now the reader has been conducted from the top of the valley to rail head and nothing remains but to wish the intending sportsman the best of luck, which includes good weather and 'tight lines'.

■ ■ ■

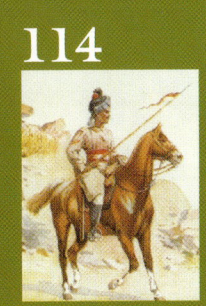

The Stinging Caterpillar
(*Euproctis icilia*)

By Major R W G Hingston, I.M.S., (Indian Army)
Year of publication: 1930

Major R W G Hingston IMS born on 17th June 1887, was commissioned in the Indian Army (IMS) on 30th July 1910. He was the medical officer to the 1924 Everest expedition. Besides the first checklist of the Birds of Dharamsala (Kangra) he authored two books A NATURALIST IN THE HIMALAYAS and A NATURALIST IN HINDOOSTAN, and wrote several articles on insects.

Hingston is believed to be the first naturalist to have seen and recorded the young of the Ibis-Bill in the wild, in a stream bed not far from the Everest Base Camp. After Everest he was stationed at Dharamsala where he compiled the first check-list of birds which was later used by Hugh Whistler.

Fabre, in his narrative of the Arbutus Caterpillar, tells of the dread which the villagers have of it. The wood-cutters, the faggot-binders, the brushwood-gatherers all join in the same tale that the caterpillar attacks with great severity, so much so that they spend a night in torment, tossing as if on live coals. The great observer seems a little doubtful of their talk. 'Do they exaggerate,' he asks. He frequently handles the caterpillars himself. He puts them on the tenderest parts of his body, but he never suffers the slightest inconvenience. He applies them to the delicate skin of his child. The result is the merest superficial irritation, nothing of the torment which the wood-cutters describe.

Here, then we have a little problem. Are these caterpillars really terrible tormentors, or have they been falsely accused? Let us turn to one of our Indian species and examine the actual facts.

One evening in July, while walking in my garden, I happen to come across a cluster of caterpillars belonging to this so-called urticating or stinging group. Science tells us that their technical name is *Euproctis icilia* Stoll. I find them on the bark of a Phalsa tree. They are quite inconspicuous, though clothed in long hairs, a mottled grey and brown covering which blends well with the underlying bark. They are worth investigation for two reasons; one, to observe their manner of development, and the other to test the problem before us. Are they venomously armed or not?

DEVELOPMENT

First, we consider their mode of development. I imprison the cluster in a glass-covered box. The caterpillars, being large, soon pupate, and in two weeks I have the adult moths. They are quiet creatures, seem as content as did the caterpillars with their prison, and make no attempt to fight through the glass. They seek no food for they have no feeding mouth-parts. All they want to do is to sit still and wait for the one event of life. This takes place a day or two after their emergence. I find them then in the act of mating, a prolonged affair, lasting many hours, with no show of enthusiasm about it, and followed, soon after separation, by a profuse discharge of eggs.

These are laid in a cluster on some suitable support. When the business is over, they make a yellow heap wrapped in a quantity of minute hairs. The source of these hairs is very interesting. They are stripped by the mother from the tip of her abdomen. There she possesses a specialized tuft from which she manages to pluck the wrap. The hairs are yellow, thin and delicate. Round the eggs they form a downy vestment, and so carefully is this covering made that not only do the hairs enclose the cluster but they also spread into its interstices and form a separate capsule for each egg. There is no viscid material in the cluster. All the eggs are kept together by the interweaving of individual hairs. Each egg in the composite mass possesses its individual nest. The eggs are not arranged in any special manner. In one place the cluster is a flat layer, in another an irregular heap. When divested of the wrappings, the eggs are almost colourless. Small hard spheres with a smooth, delicately pitted surface, they glisten with a faint lustre like a heap of tiny beads.

The eggs hatch on the sixth day. The new-born caterpillar is, of course minute. Four hours after birth it measures one-seventh of an inch. It is pale in colour, with a brown head and a number of dark spots on its back. Its whole body bristles with hairs, not just a sprinkling as in many young caterpillars, but well developed radiating tufts springing from its back and sides. Its body shows a few faint markings. Behind the head is a brown patch with a yellow spot on each side of it. There is also a yellow blotch on the tail, and two raised brown spots on the back which later will support specialized hairs.

When first born, the caterpillars are very active. They begin life by making explorations, climbing over the heap from which they have escaped, thrusting their heads into the fluffy mass, scattering the hairs in all directions, rooting into the interstices of the hairy capsule in order to get at the egg-shells inside. This is their first attempt at feeding, a combined effort of the whole family to engulf the egg-shells from which they emerged. Succeeding in this, their instincts change and they commence to nibble at a leaf. Of course their first efforts are very feeble. All they do is to tear off the finest layer of epidermis, leaving behind some yellow specks showing where this cuticle has been eaten away.

The next morning I find the family more scattered. The hairy heap has been dug out, the egg-shells have gone, nothing but a scattering of fluff remains. The caterpillars are all over the leaf. Their colour has grown a little darker and their body markings are more distinct. The problem arises, how are we to feed them? The cluster I had found was on

Lymantrid Caterpillar

SHUBHALAXMI VAYLURE

Lymantrid Caterpillar

the trunk of a Phalsa tree, so first I try them with Phalsa leaves. But these will not suit their particular taste, nor will leaves of the Imli, the Ber, the Rose, all of which grow in the vicinity of where the original cluster was found. I feel sure that the family will die of starvation, when by mere chance, in these gastronomic testings, I happen to give them a sprig of Babul, *Acacia arabica*. This evidently suits their taste. They soon climb on it, get in between the leaves, attack their edges, excavate them semicircularly, and end by stripping them completely from the stalk. Thus is the family saved from starvation; and, no doubt, under the natural conditions, this acacia is their accustomed food.

By the third day they are quarter of an inch long. Their bristles are already showing signs of specialization. Each is no longer a plain straight hair. It has changed into a central acute pointed shaft with an encircling armoury of spines. This is the foundation of that delicate machinery which becomes of such importance at a later date. On this day or the next the skin is shed. A rent occurs behind the head, and the caterpillar, little changed in appearance, comes out in a fresh coat. Every fragment of the integument is changed, even the delicate microscopic barbules have their thin coverings renewed. Later we shall see how complex are these hairs, and will realize how perfect is the act of desquamation which permits these hairs, without the breaking of a barbule, being drawn from their intricate sheaths.

The sixth day shows the caterpillar a little larger. It now begins the dropping dodge, falling, when disturbed, on an invisible thread and swinging suspended from its leaf. Also it is getting more attractive in appearance; its pigment is collecting into longitudinal streaks and its early coat of tufted bristles is changing into silky hairs. With undulating movements it crawls about the foliage, having only two objects of any importance, feeding and casting its skin. A dangerous feature is the development of cannibalism. One by one the members of the family disappear. Some happen to be smaller, more undeveloped than others, and these less fortunate individuals are devoured by the monsters that grow at their expense. The loss of life in this way becomes really serious. On the forty-fifth day only four survive out of the family of twenty-six.

The seventh week finds them fully developed. They are brown in colour, mottled with grey, about one and a half inches long. A fringe of long hairs decorates each side. From the back projects a similar armoury, but the hairs are there shorter and thicker and are ornamented with snow-white spots. As is usual in this family, Lymantriidae, the caterpillar possesses some specialized hair-tufts. In this species there is a double pair on the back at the junction of the front and middle thirds of its body, also another similar pair just a little in front of its tail.

Now we come to the pupal change. The caterpillar gets into the nearest debris, spins about its body an open network, nothing very skilful about it, just a mere reticulation of threads. Safely inside, it undergoes a shrinkage. Its hairs drop off and leave a bare skin which soon hardens into a kind of case. In this way it lies dormant for about two weeks. Then the case bursts, and out comes a small and fluffy moth. The apparition is a little surprising, the moth is such an insignificant production compared with the caterpillar that gave it birth.

URTICATION

If the reader has persisted through these tedious details, it may, perhaps, be worth his while to examine the caterpillar's poison-discharge. I recall a day when, ignorant of this insect, I happen to take one in my hand. Being interested in the defences put up by these creatures, I drop it in the midst of a swarm of ants in order to see how it wards them off. Nothing worth mention happens in the swarm, but soon I begin to feel an itching. It commences on the neck, just a slight irritation, nothing more than a mosquito-bite. I scratch it. The discomfort increases. The tingling spreads round the neck and down along the front of the chest. Then the forearms begin to smart. Clearly this is no local irritation, but something with a wider effect.

I examine the tingling patch. An intense erythema is rapidly developing. The skin is bright red and beginning to swell. There is a central inflamed spot and round it a crop of angry points. Fresh crops appear in different places. Soon the whole chest and abdomen are involved. New eruptions develop on the legs and back, and soon the whole body burns madly with one incessant itch. It becomes impossible to refrain from tearing at it, though this, of course, intensifies the torture and drives the poison still deeper into the flesh. I am almost beside myself with rage and irritation. Fortunately there is a river handy. Throwing off my clothes, I rush into the water. Some relief is given, only a little, for the painful burning still continues and at times the itching grows intense. Almost every part of the body is involved. The hands and face have escaped from the thickness of their skin, and for some reason the middle of the back is immune. The original red patches have grown into lumps, painful whitish weals.

In half an hour the main intensity of the inflammation lessens. I dread to put on my clothes again. They must certainly be infected with the poison hairs, and the torment will commence anew. However, there is no alternative. On they go,

and I rush for home. The distance is no more than half a mile, but I have not gone a hundred yards when the mad irritation breaks out afresh. Scratching and tearing at it, I again throw off my clothes and plunge into a cold bath. This again provides some temporary alleviation. Methylated spirit gives the final remedy. I rub it thoroughly all over my body, and in fifteen minutes the irritation subsides.

One could scarcely believe that a single caterpillar would be capable of causing such unendurable discomfort. How subtle must be the venom, and how marvellous is the creature's capacity of spreading its poison far and wide! An attempt to relieve it only scatters it farther until the whole body gets involved. I keep up the application

Arctid Caterpillar

of spirit to the patches. Yet, in spite of it, the urticaria persists for days. Patches of redness remain in the skin. Fresh outbursts of irritation develop and require more spirit to keep them down. Even after four days of treatment, it is necessary to apply the spirit at night. How does this simple remedy act? I suppose by neutralizing or destroying some poison. How, otherwise, could it give such immediate relief? The torment, in the absence of this remedy, would be terrible. Then could the wood-cutters truly state that they passed their nights 'tossing and turning as though on a bed of live coals'.

Surely we have here a magnificent defence. A poison which will drive a man to madness must be of value against many foes. But I foresee what might be thought to be a serious objection. The poison is distinctly delayed in its action, therefore the caterpillar will have been destroyed by its enemy long before the enemy will feel painful effects. How then can the painful effects of the poison in any way protect the caterpillar? Only, I take it, through its educating influence. Only by birds and other enemies learning through experience in the same way as do the woodcutters that these creatures are dangerous to touch. Birds without doubt do learn in this way and the highest authorities could be quoted to show that the whole principle of warning devices depends on the enemy's capacity to learn. A bird is not afraid of a warning colour through any instinctive fear of the warning but just because it has learnt its lesson as a child learns to dread the fire. I think it is exactly the same in the case of these uricating caterpillars. They escape because their enemies have learnt what they are.

Thus we see a succession of processes. First comes the extraction of the hair then the breaking of the hair into spines then the entering of the spines into the skin then the liberation of some subtle poison followed by an inflamed patch. The fingers are then brought to the infected spot. Some spines stick to them. The spines are carried by the fingers to other parts of the body. New foci of inflammation are started. These again are scratched. More and more foci are formed in the same way until in the end the whole body is involved. Thus we see in what a simple mechanical manner can the poison of a small caterpillar produce such a wide-spread effect.

The possession of poison-hairs is fairly wide-spread amongst caterpillars. For some strange reason, it seems to be confined to the caterpillars of moths. At least, I know of no butterfly larva armed in this elaborate way. The LASIOCAMPIDAE, EUPTEROTIDAE, LIMACODIDAE, ARCTIIDAE, are the chief families of Indian moths whose caterpillars have evolved the poison hair.

Crambid Caterpillar

SHUBHALAXMI VAYLURE

PURPOSE OF THE SIMPLE HAIR

We have seen the purpose of the poison hair. But this is a highly specialized contrivance. What about that far more common structure, the plain, simple, ordinary hair in which such myriads of caterpillars are clothed? What is its function? Why should such a crowd of caterpillars go in for the silky dress?

Different suggestions have been put forward. I shall mention them only briefly, and leave to the end the personal views which I have come to from observation in the field.

1. **Use as a breakfall.**—This suggestion I believe to be quite fantastic. It is that the caterpillar's hairy coat of hairs projecting from it makes a kind of springy garment which lessens the shock when it strikes the earth. Certainly hairy caterpillars do fall to ground. Many kinds, when alarmed, roll themselves into a coil and allow themselves, to drop from a twig or leaf. But they cannot need this specialized buffer, for think of the numbers of naked caterpillars which habitually drop to earth and never come to any harm. It is not a bump that a caterpillar fears, but a stab from the ovipositor of a parasite or a pinch in the beak of a bird.

SHUBHALAXMI VAYLURE

Lymantrid Caterpillar

2. **Use as a slip-away device.**—A number of hairy caterpillars, when alarmed, roll themselves up into close-wound coils. This attitude is said to protect them by preventing birds from getting a grip. A caterpillar, for instance, of the Great Tiger Moth, when rolled up, is not easy to take hold of. The covering of hairs makes it slip through the fingers, and it may be that for the same reason it will slip from the beak of a bird. In that fine old work of Kirby and Spence we are told that the larva of *Anthrenus musorum*, a destructive beetle that gets into cabinets, is covered with numbers of diverging hairs which cause it to glide from between the fingers as if it were lubricated with oil.

Thus it may be in some special instance, though I have little faith in the idea, that the hairs may function as a slip-away device.

3. **Use by being distasteful to birds.**—This view brings us to surer ground. It is a fairly well-established fact that birds and lizards do not like hairy caterpillars. The hairs, no doubt, are mechanically unpleasant to them. Also it is obvious to common observation that many kinds of hairy caterpillars travel quite fearlessly in the open, careless of any attack from birds. Nevertheless it is well known that cuckoos will eat quantities of hairy caterpillars. They seem to have some capacity for dealing with the problem which is not possessed by other kinds of birds. This little point would be worth investigation. Does the cuckoo digest the hairs, or does it vomit them forth again as birds of prey do feathers and bones?

4. **Use as a decoy**.—Certain caterpillars have their hairs collected into tufts, conspicuous projecting brushes which immediately attract the eye. In these instances it is suggested by Professor Poulton that the tuft is a kind of decoy. A bird or lizard, making a grab at the caterpillar, will seize hold of the conspicuous tuft. The hairs, which, in such cases, are loosely attached, will come out as soon as they are grabbed and the bird will get not a luscious caterpillar, but a mouthful of irritating hairs. The chief thing about this explanation is that it has been proved by means of experiment. Professor Poulton gave these caterpillars to lizards in a cage. They did grab the tufts and the hairs came out. Also the lizards were discomfited and refused to repeat the attack.

5. **Use as a frightening device**.—Caterpillars belonging to the Lasiocompid family best illustrate this peculiar use. On their backs, a short distance behind the head, are two transverse slits. When the caterpillar is touched these slits open and there is suddenly thrust out from them a black brush of velvet hairs. It is an abrupt and astonishing performance. One would think that the skin had burst asunder to let out these conspicuous brushes. Their extrusion is so unexpected that one is just a little frightened at the apparition and feels that with this creature one had better not interfere. For a little while the tufts remain protruding; then they are slowly drawn back and hidden once more within the slit.

The sudden protrusion of these dorsal brushes is clearly for the purpose of causing alarm. It is a device having the same object as the assumption of a face apparition by the puss caterpillar, or the snake-like appearance by the caterpillars of some Sphingidae; a sudden unexpected frightening manoeuvre to intimidate a lizard or an insectivorous bird.

6. **Use as a protection against ants**.—Ants are amongst the most serious enemies of caterpillars, at least on the plains of India. Naked caterpillars are completely at their mercy, but the hairy ones are practically immune. Indeed we may advance the general rule that if a caterpillar happens to be naked, then it has some special protective device. It goes in for colour assimilation or alarming attitudes, or snake-like resemblance, or offensive discharges, or horn-like protuberances which it lashes about. But when the caterpillar is covered with hairs, it discards the use of these extravagances. Its hairs are protection enough.

Now see how this works with the ants. The caterpillar of *Thiacidas postica* is very suitable for investigation. It is common, clothed in long hairs, and feeds on the Ber which is frequented by black ants. These ants, *Camponotus compressus*, are particularly fond of caterpillars. They destroy hairless ones wherever they find them, but they never touch the hairy *Thiacidas*. The caterpillar has no special capacity for resistance. When disturbed, all it does is to roll into a coil and allow itself to drop to the ground. Its immunity from the ants depends on its hair. I place one at the entrance to an ants' nest. But the ants will not touch it. They approach it, threaten it with wide-open jaws; obviously they would like to pitch it aside, but they fear the shield of hairs. Occasionally a worker may take hold of a hair, but the hairs happen to be only loosely attached, and all that happens is that the hair comes out. As a rule, the slightest touch of a hair is sufficient to repel an ant. One thing is perfectly clear. The garment of hairs gives complete immunity. The ants will not dare to penetrate the defence in order to get at the caterpillar's skin. In the end the caterpillar crawls away unharmed leaving behind it a tumult of ants.

An experiment will convince anyone who doubts. I pluck one of these caterpillars bare, a quite simple operation, since the hairs are loosely fixed and come out with a gentle tug. To the ants I give this stripped larva. It is as active as ever and runs about, apparently none the worse for the loss of its coat. But now it is completely at the mercy of the ants. They fall on it, grip it, seize it on all sides and drag it into the nest. Thus I feel sure that, at any rate in India, one of the chief uses of the caterpillar's hair is to guard it from the swarms of ants.

7. **Use as a defence against parasites**.—The most dangerous of all caterpillar-enemies are parasitic hymenoptera and diptera. These foes possess long spear-like ovipositors which they thrust into the body of the caterpillar. Now, in order to reach the caterpillar's body, the parasite must either enter the hairy investment or else have an ovipositor of greater length than that of the caterpillar's hair. Here is an incident one may sometimes witness. A hairy caterpillar is on a leaf. A parasite approaches and comes close up to the hairy barrier which extends all round the caterpillar's body. It then tries to get at the caterpillar. It puts its ovipositor between the hairs, then feels and pushes in different directions, trying to reach the caterpillar's skin. It will not itself enter the barrier. Like the ants, it dreads the touch of a hair. Its success depends on whether its ovipositor is long enough to reach the caterpillar's skin. If not, the parasite goes off.

SHUBHALAXMI VAYLURE

Monkey moth caterpillar

This little observation from India is confirmed by something similar from Africa. Dr. Carpenter was watching some hairy caterpillars belonging to the family Eupterotidae. He saw a Tachinid fly approach them. It sidled about them in an amusing manner, but all the time kept facing the caterpillars. Now there was only one opening in the fence of hairs. The head end of the caterpillar was comparatively bare, and the fly seemed to know this point of weakness, for it made its attack at this open spot. It elevated itself on its hind legs, pushed underneath itself an enormous ovipositor the tip of which stuck out in front of its head. By this means it managed to place an egg in the open space near the head of the caterpillar. If it could not get at the head, then it tried to push its ovipositor between the hairs. But it clearly chose the open spot because it disliked the caterpillar's coat.

These little observations, I believe, give us the main clue to the use of the caterpillar's hairy covering. It is not an uncommon arrangement for the hair to be distributed in the form of a fringe along either side of the caterpillar. Take for example the Lasiocampids. In these the fringes are so thick that the caterpillar is surrounded as if by a fence. For what purpose? Because the flank attack is the most dangerous. Hymenopterous parasites dislike hairs, and, so far as I have seen, do not alight direct on the caterpillar's body. They first settle on the leaf, then creep towards the caterpillar. Hence the danger to the caterpillar is from the sides. What it particularly wants is a fringe that surrounds it like a ringed fence.

We know so little, indeed scarcely anything, of the way in which these parasites perforate their victims. For example, what parts of the caterpillar are vulnerable? What parts do the parasites habitually make for? What is the length of the parasite's ovipositor, and its relation to the length of the caterpillar's hair? These and a hundred other questions will have first to be answered before we can know the whole reason for the arrangements of caterpillar's hairs.

Certainly we have a fragment of knowledge. Here and there we have a record of a particular parasite being the enemy of a particular caterpillar. But what do we know beyond the names of the species? Where are the necessary field observations that will help us to elucidate the point in question? They are almost altogether absent. Too much care and patience is necessary for Naturalists even to dream of such a task. I believe if we had carefully detailed records of the manner in which say fifty parasites tried to penetrate fifty kinds of hairy caterpillars, the result would throw a flood of light on the multifarious designs of caterpillar structure. We should learn much that we never suspect of the uses of many of those peculiar appendages of spines, tufts, hairs, nodules, horns, tails, brushes, tussocks, and other strange superficial structures for which at present we see no use.

■ ■ ■

The Asian Elephant

By Lieut Col J H Williams, O.B.E. (Indian Army)
(Elephant Bill)

Year of publication: 1950

Lieut Col J H Williams, after gaining experience of camels and mules in World War I, joined the staff of the Bombay Burma Trading Corporation in the teak forests of Burma. After some twenty years' close acquaintance with elephants both tame and wild in that country, he was given war commission as a Lieut on 09 Nov. 1942 in the Royal Indian Army Service Corps (affiliated to the Madras Regiment) and became "Elephant Adviser" to the XIVth Army when the Japanese had overrun Burma in World War II. The result of these twenty-five or so years' intimate connection with Burmese 'oozies' and their elephants has resulted in an intimate and first-hand account of elephants in his book ELEPHANT BILL, first published in 1950.

The fact that men and elephants live about as long as one another and come to maturity at much the same ages means that they can live together all their lives. They can thus acquire a lifelong mutual knowledge of each other's characters. With no other domestic animal is this possible. A baby boy may be born in an elephant camp, and at the same moment an elephant calf may be being born a mile or two away in the jungle; and that child and that calf may grow up together, play together, work together all their working lives, and they may still be familiar friends when sixty years have passed.

Elephants are not bred in captivity. The captive animals breed naturally in their natural surroundings. During the war I was talking about elephants to two war correspondents, one American and the other an Australian. The latter asked me: "Is it true that elephants are very shy about their actual love affairs?" Before I could answer, the American chipped in with: "Of course they are: aren't you?" The mating of elephants is a private affair, and even the oozies of the tusker and the female concerned may not know that it has taken place. Often they know, but regard it as none of their business, and do not talk about it.

The most fantastic tales are told, and even believed, about the mating habits of elephants by Europeans. The tallest story that has come my way was told by a young Sapper officer to a very attractive nurse whom he took to the Rangoon Zoo, after the recapture of Rangoon in May, 1945. Among the few animals left behind in it by the Japs were two young elephants, a male and a female, which led very boring existences, hobbled and tethered by the hind legs to two posts in the elephant-shed. While watching this melancholy pair, the young Sapper described how the female elephant turns to thoughts of love in the spring time, and prepares for her honeymoon by digging a deep pit, round which she stacks a month's supply of fruit and fodder for herself and her young bridegroom. When she has completed these preparations, she lies down in the nuptial pit and trumpets a love-call to her mate. After his arrival they live in one *unending embrace* for the whole month, and do not separate until they have shared their last pineapple or banana! No doubt the Sapper hoped that his attractive listener would take the hint and act likewise. The pretty girl to whom this story had been told afterwards applied to me for confirmation of the story. I felt sorry to have to disillusion her, though the love-making of elephants as I have seen it seems to me more simple and more lovely than any myth.

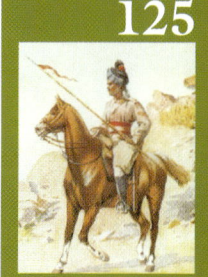

It is beautiful because it is quite without the brutishness and the cruelty which one sees in the mating of so many animals.

Without there being any appearance of season, two animals become attracted by each other. In other words they fall in love, and days, and even weeks, of courtship may take place, the male mounting the female with ease and grace and remaining in that position for three or four minutes. Eventually the mating is consummated, and the act lasts five or ten minutes, and may be repeated three or four times during the twenty-four hours. The pair will keep together as they graze for months, and their honeymoon will last all that time. When they have knocked off from the day's work they will call each other and go off together into the jungle. My own belief is that it lasts until the female has been pregnant for ten months — that is, until she has become aware that she is pregnant. The act of mating can be performed as easily by an elephant wearing hobbles as without them as the position of the male's forelegs lying along the barrel of the back of the female is not interfered with by the hobbles. In the final mating position the male is standing almost vertically upright, with the forefeet resting gently on the female's hindquarters.

NIKHIL BHOPALE

The average female first mates between the ages of seventeen and twenty. She shows no sign of any particular season, but apparently feels some natural urge. It has recently been noticed that female circus elephants become moody in periods of approximately twenty-two months. Gestation lasts twenty-two months, and she does not appear to realise that she is pregnant until the end of the first ten months. After that the period of mating comes to an end, and the companionship of the male is replaced by that of a female friend or "auntie". From that time onwards the expectant mother and her girl friend, or "auntie", are never apart. They graze together always, and it becomes difficult to separate them. It is, indeed, cruel to do so. Their association is founded on mutual aid among animals, the instinctive knowledge that it takes two mothers to protect a calf elephant against tigers, which, in spite of all precautions, still kill twenty-five per cent of all calves born.

ASHOK KUMAR

After the calf has been born, the mother and the "auntie" always keep it between them as they graze—all through the night—and, while it is very young, during daylight hours as well.

To kill the calf the tiger has to drive off both the mother-and "auntie" by stampeding them. To do this he will first attack the mother, springing on her back and stampeding her; then he returns to attack "auntie," who defends the calf, knowing that in a few moments the mother will return. On many occasions I have had to dress the lacerated wounds of tiger-claws on the backs of both a mother elephant and her friend.

A mother elephant in captivity has no suspicions that man will injure her calf. I have only once been attacked by one of the many mothers whom I have congratulated by a pat on the trunk, often within an hour of the actual birth. That was an accident. I was patting a calf so young that it could not focus me with its little piggy eyes, and it bumped hard

against my bare knees and yelled out the cry for danger. As I jumped back, the lash of the mother's trunk missed me by inches. She then chased me, but only for twenty yards, as she had to return to her squealing babe. On the other hand, I have handled a newly born dwarf elephant under the mother's belly, lifting it up so that it stood on a small platform, so as to reach its mother's nipples, and the mother seemed to consider this as much of a joke as I did.

A baby calf follows its mother at heel for three or four years. It is suckled by the mother for that period, from the breasts between her forelegs.

This position, between the forelegs, affords the calf perfect protection. At birth the calf's trunk is a useless object, or membrane, growing rather to one side, so as to allow the calf to suck more easily through the mouth. It does not become flexible and useful for three to four months. When the sacred white elephant of Mandalay Palace was a calf its mother died, and it was suckled by twenty young Burmese women daily as wet nurses, and so reared.

At the age of five or, at most, six years, the calf has learned to gather its own fodder, and gradually gives up sucking its mother. Female elephants have an average of four calves in their life-time. Twins are not uncommon, and two calves of different ages following their mother at heel is quite a usual sight. Larger families are not uncommon.

"Accidents" happen in the elephant world, and elderly females occasionally spring a surprise. Thus Main Hpo (a Shan name) gave birth to her eighth calf just after the Japanese War, in the Gangaw Valley, when she was sixty-one years old. All her previous calves had lived and had been trained–in fact her eldest son, a tuskless male named Hine Pau Zone (Mr Laziest) was the only elephant recorded as having killed a Japanese soldier. This incident took place at a Chin Village in 1945, during the Japanese withdrawal.

After weaning, young elephants go through an awkward stage, becoming a bit truculent owing to the desire for independence–much like human boys and girls.

ASHOK KUMAR

ASHOK KUMAR

At fifteen or sixteen they become very much like human flappers and young stalwarts. They have reached the same adolescent stage of not knowing quite what they want. Some soon find out, others do not; for their temperaments vary.

Young male elephants do a lot of flirting with the females from the ages of sixteen to twenty, some being most enterprising. But the average animal does not show any signs of musth until the age of twenty. A male elephant will mate when he is not on musth, in fact he usually does. But when he is on musth all the savage lust and combative instincts of his huge body come out.

From the age of twenty to thirty-five musth is shown by a slight discharge of a strongly smelling fluid from the musth-glands near the eye, directly above the line of the mouth. In a perfectly fit male it occurs annually during the hot months, which are the mating season. It may last about two weeks, during which time he is very temperamental.

From the age of thirty-five to forty-five the discharge increases and runs freely, eventually dribbling into his mouth, and the taste of it exasperates him and makes him much more ferocious. He is physically in his prime at that age, and unless he is securely chained to a large tree while on musth, he is a danger to his oozie and to other elephants. His brain goes wild, as though nothing would satisfy him, and nothing will.

From forty-five to fifty musth gradually subsides, and finally disappears. Tuskers that have killed as many as nine men between the ages of thirty-five and forty-five will become docile during musth in the later years of their lives. But no elephant on musth can be trusted unless he is over sixty years old.

Poo Ban, a magnificent tusker, was normally a friendly animal, and would allow me to walk under his head and tusks, but he went on musth in the Taungdwin Forest area, killed his oozie and another man, then killed two female elephants, and attacked on sight

The author with Molly Mia, one of his nineteen dogs

any man who came near him. Finally he entered villages, tore rice granaries open, and became the terror of the valley. I offered a reward of three hundred rupees for his capture, and decided to destroy him if he could not be captured.

He was marked down in a dense patch of bamboo jungle in Saiyawah (the Valley of Ten Villages), four marches away. With Kya Sine, my gun-boy, I set out, lightly loaded with two travelling elephants as pack. The evening before I was to tackle Poo Ban I was testing my rifle with a half charge, i.e., with half the cordite removed, and with a soft-nosed bullet in the left barrel, keeping a normal hard-nosed cartridge in the right. I wanted to wound Poo Ban in one of his forefeet with the half charge, and then recapture him, break his spirit and heal his wound. At a hundred yards my practice-shooting was so accurate that I felt hopeful of success. Kya Sine, however, begged and implored me to let him go ahead and attempt to recapture him without shooting, so that he could earn the three hundred rupees. He intended to tackle him boldly, face to face, relying on his own authority and the animal's habits of obedience. Unfortunately, I gave in, and before dawn he had gone on ahead. I arrived at three p.m. next day to be met by men who said: "Kya Sine is dead." Poo Ban had killed him during his attempt at recapture.

That night I bivouacked in an open place which had at one time been paddy-fields. It was a brilliant moonlight night, and before I went to sleep I made my plans to recapture Poo Ban. I had no desire to avenge the death of Kya Sine—to whom I was devoted, and who had the greatest knowledge of jungle lore of any man I have ever known. The idea of revenge on an elephant would have been very distasteful to him.

I was asleep, lying in the open, when I was woken by a clank! clank! clank! Luckily for me, a piece of chain had been left on Poo Ban's forefoot. I came suddenly to my senses out of a dream, and, jumping up, saw the finest sight of my life. Two hundred yards away, in the open, a magnificent tusker was standing, with his head erect in challenge, defiant of the whole world. He was a perfect silhouette. I did not dare move an eyelid. He was a bigger man than I was, and while I held my breath he moved on with a clank, clank, clank, which at last faded away like the far sound of the pipes over the hills.

At dawn I tried to put my plans into action. When he had been located, I took up my position, while twenty Burmans, with four shot-guns between them, tried to drive him past me.

Poo Ban faced the lot, defying them four times. Shots rang out, but at last he changed his mind and, turning, came towards me. I was perched on a broken down, dilapidated brick pagoda, a heap of rubble about six feet high, behind which was the hundred-feet-high bank of the Patolone River. Directly in front of me was a clearing of disused paddy-fields, and my hopes were that Poo Ban would cross it. I still meant to wound him in the foot, recapture him, break his spirit and then heal the wound.

Poo Ban came out of the jungle with his head held high. He halted, and then made a bee-line across my front, travelling, fast over the open ground.

Kneeling, I took the shot at his foot on which my plans depended. The bullet kicked up a puff of dust in front of his near forefoot as he put it down in his stride. I had missed!

Poo Ban halted and swung round to face me, or the bark of my rifle which he had heard. Then he took up the never-to-be-forgotten attitude of an elephant about to charge, with the trunk well tucked away in his mouth, like a wound-up watch-spring or the proboscis of a butterfly. As he charged it flashed through my mind that I had no time to reload. I depended on the hard-nosed bullet in the right barrel. At twenty-five yards his head dropped; his tusks drove nine inches into the ground. For a few seconds he balanced, and then toppled over, dead.

I dropped my rifle and was sick, vomiting with fear, excitement and regret. Poo Ban was dead, and I had failed to catch him alive. There was no court of inquiry. My report was accepted, and I was given the tusks as a souvenir, a souvenir of a double failure that I bitterly regretted, and of the death of the finest and bravest Burman hunter I have known.

From A.D. 1024 onwards elephants are mentioned in trains of thousands in the Wars of the Princes; and those who are familiar with the subject find no reason to question such figures. After the battle of Delhi, Prince Timour is stated to have captured three thousand elephants from Prince Mohammed. It is said that they all had snuff put into their eyes so as to make them appear to weep tears of grief at having been defeated.

Indian elephants were on the strength of the Royal Engineers up till 1895, when Daisy, the last and oldest pensioner, died.

In India itself elephants have gradually disappeared. Only a few are kept for ceremonial purposes. The elephant is, however, an animal in which every Indian is interested, and it is invested in a haze of myth and legend which delights children and is a source of pride to the descendants of India's ancient warriors.

It is impossible to understand much about tame elephants unless one knows a great deal about the habits of wild ones.

Wild elephants normally live in herds of thirty to fifty, and during the year cover great distances, chiefly in search of fodder. During the monsoon months–from June to October–they graze on bamboo in the hilly forest country, sometimes remaining on one watershed for a week or ten days, after which they suddenly move ten miles for another week's stay on another slope. After the monsoons are over they move into the lower foothills and the swamp valleys, feeding more on grass and less on bamboo.

It is at this time that the full-grown male tuskers join the herd, though they seldom actually enter it, preferring to remain on its outskirts, within half a mile or a mile of it. At this season they do their courting and mating, in the course of which the older bulls often have to fight some youngster who is pursuing the same female.

The herds know their yearly cycle of grazing grounds, and in their annual passage wear well-defined tracks along the ridges of the hills. In places where they have to descend from a precipitous ridge down the side of a watershed they will move in Indian file; and by long use will wear the track into a succession of well-defined steps.

Wild elephants hate being disturbed on their feeding-grounds, but they do not usually stampede suddenly, like many other herds of big game. With an uncanny intelligence, they close up round one animal as though they were drilled, and their leader then decides on the best line of retreat. She leads, and they follow irresistibly, smashing through everything, like so many steamrollers.

If they cannot exactly locate the danger which threatens them, they invariably retreat along the track from which they have come while grazing, with their trunks on each other's backs, but in a formation of three or four abreast.

I once had the unpleasant, but exciting, experience of being a member of such a stampeding party, when I was mounted on one of my own elephants. The wild elephants were fortunately quite oblivious of the fact that Elephant Bill and tusker Po Sein (Mr Firefly) were among them. Fortunately my rider was able to extricate us from the party before they reached a muddy nullah with banks eight feet high. The leading elephants plunged their forefeet into the edge of the bank, broke it away, and, sitting on their haunches, made a toboggan slide for the herd following them.

Most wild elephant calves are born between March and May. I believe that, if she is disturbed, the mother elephant will carry her calf, during its first month, holding it wrapped in her trunk. I have seen a mother pick up her calf in this way. On two occasions

I have found the tracks of a newly born baby calf in a herd. Later on, after I had disturbed them, there were no tracks of the calf to be found among those of the stampeding animals, nor could the calves have kept up with a stampeding herd. But there was no possibility of the calf being hidden or abandoned.

The birth of a calf is quite a family event in a herd of wild elephants, and I have on several occasions, camped close to what I may call the maternity ward. For many years I could not understand the bellowing and trumpeting of wild elephants at night during the hot weather, when most calves are born. The fuss is, without any doubt, made by the herd in order to protect the mother and calf from intruders – in particular from tigers. The noise is terrifying. The herd will remain in the neighbourhood of the maternity ward for some weeks, until the new arrival can keep up with the pace of a grazing herd. The ward may cover an area of a square mile, and during the day the herd will graze all over it, surrounding the mother and her newly born calf, and closing their ranks round her at night. The places chosen, which I have examined after the herd has moved on, have been on low ground where a river has suddenly changed its course and taken a hair-pin bend. These spots were thus bounded on three sides by banks and river. The kind of jungle found in such places is always the same. They are flooded during the rains, but during the hot weather–the normal calving period–they are fairly dry, with areas of dense kaing, or elephant grass, eight to twelve feet high, with an occasional wild cotton-tree giving shade. They are eerie spots, and to explore them is an adventure. Wild pig breed in the same type of jungle, and harbour their sounders of suckling pigs under huge heaps of leaves and grass which in size and appearance resemble ant-heaps four feet high.

It is common practice for a Burman oozie, or elephant-rider, to ride his elephant silently up to such a "pig's nest" of leaves and grass and then, silently controlling the elephant by movements of his foot and leg, to instruct him to put one forefoot gently on the mound. Squeals and snorts usually follow from the old sow, and three or four suckling pigs join her in a stampede.

It is a peculiar thing that the elephant, which becomes so accustomed to man, and has such confidence in him once it has been trained, should be so afraid of him in its wild state. Owing to this fear of man, they do surprisingly little damage to village crops, considering the vast numbers of wild elephant. They much prefer their own deep jungles, and seldom leave them. The damage that they do has been greatly and most unfairly exaggerated, and the extermination of wild elephants in Upper Burma was actually started

ASHOK KUMAR

on unreliable advice. Solitary animals may, however, do great damage and become bold enough to drive off any human intruder who shows himself. They will do this almost as though they thought it was a joke. Such animals, however, are always eventually declared rogues and are killed–or at least shot at, or caught, or injured in traps.

Herds of wild elephants are not always suspicious of danger. I have on many occasions ridden on one of my own tuskers into a herd of sometimes as many as fifty animals. Sitting on my own elephant, I have passed so close to a wild one that I could have struck a match on his back. Without being detected, I have watched and photographed wild calves of different ages playing in a mud wallow, like children playing at mud pies.

The mating of wild elephants is very private. The bull remains, as usual, outside the herd, and his lady love comes out where she knows she will find him.

She gives the herd the slip in the evening, and is back with them at dawn. Sometimes a rival tusker intervenes, and a duel ensues. This is why elephant-fights are always between two bulls. There is never a general dog-fight within the herd.

Elephant bulls fight head to head and seldom fight to the death, without one trying to break away. The one that breaks away frequently receives a wound which proves mortal. Directly one of the contestants tries to break off and turn, he exposes the most vulnerable part of the body. The deadly blow is a thrust of one tusk between the hind legs into the loins and intestines where the testicles are carried inside the body. It is a common wound to have to treat after a wild tusker has attacked a domesticated one.

Some males never grow tusks, but these tuskless males are at no disadvantage in a fight, although to outward appearances they are the eunuchs of the herd. This impression is quite wrong. From the age of three all that the animal gains by not having to grow tusks goes into additional bodily strength, particularly in the girth and weight of the trunk. As a result, the trunk becomes so strong that it will smash off an opponent's tusk as though, instead of being solid ivory, it were the dry branch of a tree.

From the time that a male calf is three years old there is always interest among the oozies as to whether it is going to be a tusker with two tusks, or a tai (with one tusk, either right or left), or a han (a tuskless male, but with two small tushes such as females carry), or, lastly, a hine, which has neither tusks nor tushes.

One of the most delightful myths about wild elephants is that the old tuskers and females drop out of the happy herd life when they realise they are no longer wanted, and that they finally retreat to die in a traditional graveyard in some inaccessible forest. This belief has its origin in the fact that dead elephants, whether tuskers or females, are so seldom found. I wish I could include a description here of how I had discovered one of these graveyards. But since I cannot, I shall try to explain away the myth by describing what really happens. I will take the case of a fine old bull that has stopped following the herd at about the age of seventy-five and has taken to a solitary existence. He has given up covering great distances in a seasonal cycle, and remains in the headwaters of a remote creek. It has become enough for him to devote all his time to grazing, resting and taking care of his health. His cheeks are sunken, his teeth worn out. Gathering his daily ration of six hundred pounds of green fodder has become too great a tax on his energy, and he knows he is losing weight. Old age and debility slowly overtake him and his big, willing heart. During the monsoon months he finds life easy. Fodder, chiefly bamboos, is easily gathered, and he stays up in the hills. As the dry season approaches, fodder becomes scarcer, and the effort of finding food greater, and he moves slowly downhill to where he can browse on the tall grass. Then, as the hot season comes on, and there are forest fires, he is too tired and too old to go in search of the varied diet he needs, and his digestion suffers. Fever sets in, as the showers of April and May chill him, and he moves to water–to where he knows he can always get a cool drink. Here, by the

ASHOK KUMAR

large pool above the gorge, there is always green fodder in abundance, for his daily picking. He is perfectly happy, but the water slowly dries, until there is only a trickle flowing from the large pool, and he spends his time standing on a spit of sand, picking up the cool sand and mud with his trunk, and spraying it over his hot, fevered body.

One sweltering hot evening in late May, when there was not a breath of air stirring in this secluded spot, to which he had come again for a drink, he could hear that a mighty storm was raging ten miles away in the hills, and he knew the rains had broken. Soon the trickle would become a raging torrent of broken brown water, carrying trees and logs and debris in its onrush. Throwing his head back, with his trunk in his mouth as he took his last drink, he grew giddy. He staggered and fell but the groan he gave was drowned by peals of thunder. He was down–never to rise again–and he died without a struggle. The tired old heart just stopped ticking.

Two porcupines got the news that night, and, in spite of the heavy rain, attacked one of his tusks, gnawing it as beavers gnaw wood.

They love the big nerve-pulp inside near the lip. They had only half eaten through the second tusk when the roar of the first tearing spate of the rains drove them off.

A five foot wall of water struck the carcase–debris piled up while the water furiously undermined and outflanked this obstruction–at last the whole mass of carcase, stones and branches moved, floated, and then, swirling and turning over, went into the gorge down a ten-foot waterfall and jammed among the boulders below. Hundreds of tons of water drove on to it, logs and boulders bruised and smashed up the body, shifting it further, and the savage water tore it apart. As the forest fires are God's spring-cleaning of the jungle, so the spates of the great rains provide burial for the dead. That elephant never had to suffer months of exhausting pilgrimage to reach a common graveyard.

By dawn the floods had subsided and the porcupines had to hunt for their second meal of tusk. Other jungles scavengers had their share of the scattered parts, taking their turns in the order of jungle precedence. But the spate came again the next night, and in a week all traces of the old tusker had disappeared.

By the time it is twenty-five years old, a well-trained elephant ought to be able to understand twenty-four separate words of command, quite apart from the signals or "foot-aids" of the rider. He ought also to be able to pick up five different things from the ground when asked. That is to say, he should pick up and pass up to his rider with his trunk a jungle dah (knife), a koon (axe), his fetter or hobble-chain, his tying-chain (for tethering him to a tree) and a stick. I have seen an intelligent elephant pick up not only a pipe that his rider had dropped, but a large lighted cheroot.

He will tighten a chain attached to a log by giving it a sharp tug with his trunk, or he will loosen it with a shake and a waggle, giving it the same motion with his trunk as that given by a human hand.

An elephant does not work mechanically, like many animals. He never stops learning, because he is always thinking. Not even a really good sheep-dog can compare with an elephant in intelligence.

I don't believe that "an elephant never forgets", but I should scarcely be surprised if he tied a knot in his trunk to remember something, if he wanted to. His little actions are always revealing an intelligence which finds impromptu solutions for new difficulties. If he cannot reach with his trunk some part of his body that itches, he doesn't always rub it against a tree; he may pick up a long stick and give himself a good scratch with that, instead. If one stick isn't long enough, he will look for another which is.

If he pulls up some grass, and it comes up by the roots with a lump of earth, he will smack it against his foot until all the earth is shaken off, or, if water is handy, he will wash it clean, before putting it into his mouth. And he will extract a pill (the size of an aspirin tablet) from a tamarind fruit the size of a cricket ball in which one has planted it, with an air of saying: "You can't kid me."

Elephants can also detach a closely clinging creeper, like ivy, from a tree far more skilfully than can a man working with two hands. This is due to their greater delicacy of touch. Many young elephants develop the naughty habit of plugging up the wooden bell they wear hung round their necks (*kalouk*) with good stodgy mud or clay, so that the clappers cannot ring, in order to steal silently into a grove of cultivated bananas at night. There they will have a whale of a time, quietly stuffing, eating not only the bunches of bananas, but the leaves and, indeed the whole tree as well. And they will do this just beside the hut occupied by the owner of the grove, without waking him or any of his family.

Catching a young animal at this is just like catching a small boy among the gooseberry bushes. For some reason stolen fruit is always sweetest.

Oozies are not always as innocent as they pretend on such occasions. I once had to pay a fine to the Forest Department for damage done by my elephants to some experimental plantations of teak saplings. Naturally, I gave the oozies a reprimand for their slackness in allowing their animals to stray into these plantations. A month afterwards I happened to meet the Forest Officer who had fined me, near a large village, where we both camped for the night. He had four elephants with him, and I had eight. Next morning his annoyance can be imagined when the village headman arrived to ask for compensation for no less than a hundred banana-trees, destroyed by his four elephants. Strangely enough, not one of my eight elephants had been involved in the mischief, a fact which made it even more annoying for him. It was not until a week after we had parted company that I found out that though my elephants were innocent, my oozies were quite the reverse. They had taken the bells off the Forest Officer's four elephants and during the night had led them quietly into the banana groves and had thus paid him out for fining me for the damage to the teak plantation.

I have personally witnessed many remarkable instances of the quick intelligence of elephants, though I cannot claim that they equal the famous yarns which delight all of us, whether we are children or grownups–such as that of the circus elephant who saw a man who had befriended him sitting in a sixpenny seat, and at once picked him up with his trunk and popped him into a three-and-sixpenny one!

But the following incidents seem to me to denote immediate brain reaction to a new situation, rather than anything founded on repetitive training.

An uncertain-tempered tusker was being loaded with kit while in the standing position. On his back was his oozie, with another Burman in the pannier, filling it with kit. Alongside, on the flank, standing on the ground, was the paijaik attendant, armed with a spear which consisted of a five-foot cane, a brightly polished spearhead at one end and a spiked ferrule at the other. Another Burman was handing gear up to the Burman in the pannier, but got into difficulties with one package and called out to the paijaik to help him. The latter thrust the ferrule of the spear into the ground so that it stood planted upright, with the spearhead in line with the elephant's eye. Then he lent a hand. The oozie, however, did not trust his beast, and said in a determined voice, "Pass me the spear." The tusker calmly put its trunk round the cane at the point of balance, and carefully passed it up to his rider. But, unthinkingly, he passed it head first, and held it as though waiting for the rider to catch hold of it by the head.

The rider yelled at his beast in Burmese: "Don't be a bloody fool, pass it right way round." With perfect calm and a rather dandified movement, the elephant revolved the spear in mid-air and, still holding it by the point of balance, passed it to his oozie, this time ferrule first.

The oozie did not say thank you, but gave him a curse with a touch of endearment– as though saying, "You are a damned ill-mannered wild elephant, and I want no more of it." Then, with a quick movement, he moved the spear-head beside the elephant's eye, an action which meant that he would suffer for it if he tried any tricks with his tusks on those engaged in loading him up. The loading was completed without incident.

Sometimes an elephant will show its intelligence by divining what its oozie wishes.

A case I remember concerned an animal which would not work with a rider on its head, but was obedient to the words of command given by its oozie walking alongside. I was watching this beast straightening logs in a creek–that is to say, placing them in rows of eight or twelve parallel to each other and pointing down the bed of the stream, in readiness for the first floods to carry them away. The oozie was sitting on the bank; work was almost finished, but, because I was around, he knew every log had to be straight in line with the others before they broke off.

There was one noticeable and unshapely log, and the elephant came to the last row in which it lay. He was a big tusker, and was doing all the work with his tusks and head, free of all chains. Without any word of command being given, he let the first log alone, and began shifting the second, keeping one eye on his oozie, as though saying: "Come on, wake up and tell me what you want!"

The oozie soon told him, shouting: "You old son of a bitch! What's wrong with that one? Leave it."

The elephant moved on to the next log, keeping his eye cocked on his oozie, like an old man looking over a pair of spectacles.

"No," shouted the oozie. "You know as well as I do," and made a gesture of picking up a stone to throw at his beast.

The elephant gave a squeal of pure delight at having pulled his oozie's leg, and, without hesitation, disregarded the next five logs and, without pausing, bent down and rolled the one irregularly placed log over four times, leaving it exactly parallel with the others and about a foot from them. Then he walked up to his master, as though to say: "Enough fooling, let's break off!" and the day's work was finished for man and beast.

But one of the most intelligent acts I ever witnessed an elephant perform did not concern its work, and might just as well have been the act of a wild animal.

One evening, when the Upper Taungdwin River was in heavy spate, I was listening and hoping to hear the boom and roar of timber coming from upstream. Directly below my camp the banks of the river were steep and rocky and twelve to fifteen feet high. About fifty yards away on the other side, the bank was made up of ledges of shale strata. Although it was already nearly dusk, by watching these ledges being successively submerged, I was trying to judge how fast the water was rising.

I was suddenly alarmed by hearing an elephant roaring as though frightened, and, looking down, I saw three or four men rushing up and down on the opposite bank in a state of great excitement. I realised at once that something was wrong, and ran down to the edge of the near bank and there saw Ma Shwe (Miss Gold) with her three-months-old calf, trapped in the fast-rising torrent. She herself was still in her depth, as the water was about six feet deep. But there was a life-and-death struggle going on. Her calf was screaming with terror and was afloat like a cork. Ma Shwe was as near to the far bank as she could get, holding her whole body against the raging and increasing torrent, and keeping the calf pressed against her massive body. Every now and then the swirling water would sweep the calf away; then, with terrific strength, she would encircle it with her trunk and pull it upstream to rest against her body again.

There was a sudden rise in the water, as if a two-foot bore had come down, and the calf was washed clean over the mother's hindquarters and was gone. She turned to chase it, like an otter after a fish, but she had travelled about fifty yards downstream and, plunging and sometimes afloat, had crossed to my side of the river, before she had caught up with it and got it back. For what seemed minutes, she pinned the calf with her head and trunk against the rocky bank. Then, with a really gigantic effort, she picked it up in her trunk and reared up until she was half standing on her hind legs, so as to be able to place it on a narrow shelf of rock, five feet above the flood level.

Having accomplished this, she fell back into the raging torrent, and she herself went away like a cork. She well knew that she would now have a fight to save her own life, as, less than three hundred yards below where she had stowed her calf in safety, there was a gorge. If she were carried down, it would be certain death. I knew, as well as she did, that there was one spot between her and the gorge where she could get up the bank, but it was on the other side from where she had put her calf. By that time, my chief interest was in the calf. It stood, tucked up, shivering and terrified on a ledge just wide enough to hold its feet. Its little, fat, protruding belly was tightly pressed against the bank.

While I was peering over at it from about eight feet above, wondering what I could do next, I heard the grandest sounds of a mother's love I can remember. Ma Shwe had crossed the river and got up the bank, and was making her way back as fast as she could, calling the whole time–a defiant roar, but to her calf it was music. The two little ears, like little maps of India, were cocked forward, listening to the only sound that mattered, the call of her mother.

Any wild schemes which had raced through my head of recovering the calf by ropes disappeared as fast as I had formed them, when I saw Ma Shwe emerge from the jungle and appear on the opposite bank. When she saw her calf, she stopped roaring and began

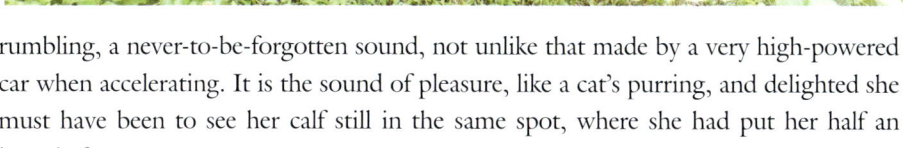

A.J.T. JOHNSINGH

rumbling, a never-to-be-forgotten sound, not unlike that made by a very high-powered car when accelerating. It is the sound of pleasure, like a cat's purring, and delighted she must have been to see her calf still in the same spot, where she had put her half an hour before.

As darkness fell, the muffled boom of floating logs hitting against each other came from upstream. A torrential rain was falling, and the river still separated the mother and her calf. I decided that I could do nothing but wait and see what happened. Twice before turning in for the night I went down to the bank and picked out the calf with my torch, but this seemed to disturb it, so I went away.

It was just as well I did, because at dawn Ma Shwe and her calf were together–both on the far bank. The spate had subsided to a mere foot of dirty-coloured water. No one in the camp had seen Ma Shwe recover her calf, but she must have lifted it down from the ledge in the same way as she had put it there.

Five years later, when the calf came to be named, the Burmans christened it Ma Yay Yee (Miss Laughing Water).

In the XIVth Army our soldiers varied in colour from white, through every shade of yellow and brown, to coal black. The animals we used reflected a similar variety. Pigeons, dogs, ponies, mules, horses, bullocks, buffaloes and elephants, they served well and faithfully. There were true bonds of affection between men and all these beasts, but the elphant held a special place in our esteem. It was not, I think, a matter of size and strength. It was the elephant's dignity and intelligence that gained our real respect. To watch an elephant building a bridge, to see the skill with which the great beast lifted the huge logs and the accuracy with which they were coaxed into position, was to realise that the trained elephant was no mere transport animal, but indeed a skilled sapper.

I could never judge myself how much of this uncanny skill was the elephant's own and how much his rider's. Obviously it was the combination of the two which produced the result, and without the brave, cheerful, patient, loyal Burmese oozie our elephant companies could not have existed. And we should have had no oozies had it not been for men like "Elephant Bill" and his assistants. It was their jungle craft, elephant sense, dogged courage, and above all the example they set, which held the Elephant Companies together under every stress that war, terrain and climate could inflict on them.

— **Quoted from Field Marshal Sir William Slim's Foreword to Col. Williams' book** *Elephant Bill*

A Popular Treatise on the Common Indian snakes

To **Col Frank Wall** we are indebted more than to any other man for our knowledge of the habits of the Indian snakes. As a member of the Indian Medical Service he arrived in the country in 1894. Here he spent most of the next 30 years of his life and in the course of his duties was stationed in most parts of the Peninsula including Sri Lanka and Burma. Wherever Wall went he collected and studied his material, and by his enthusiasm induced others to collect for him. He was not a museum worker. His interest was in the living creatures and his voluminous writings deal almost entirely with their habits and structure. His larger works include 'The Snakes of Ceylon', 'The Poisonous Snakes of our British Indian Dominions' and 'A Popular Treatise on the Common Indian Snakes.'

He saw active service in the First World War and was with the expeditionary Force to Mesopotamia. He also served in France. He was twice mentioned in dispatches and in 1915 received the C.M.G. for his work at Bologne.

THE COBRA

By Col Frank Wall, I.M.S., (Indian Army)

Year of publication: 1913

History.—As one would expect the cobra is one of the very earliest snakes to receive mention in scientific literature. Long before the inauguration of scientific nomenclature it was referred to. Seba in 1734 appears to have been the first to describe, and feature it. Later in 1754 Linné figured it, and from him it received its scientific baptism under the name *Coluber naja*. Laurenti was the next to refer to it under the name *Naja lutrescens* in 1768, and then Russell described and figured it in both his volumes which appeared in 1796 and 1801.

Nomenclature.—(a) Scientific.—The generic name introduced by Laurenti, was borrowed from the specific title bestowed upon it by his predecessor Linné. "Nag", the Sanscrit word for snake, is most probably the origin of Linnés "Naja" but why the "j" has been substituted for the "g" it is difficult understand.

The specific title "tripudians" bestowed by Merrem is from the Latin meaning literally "dancing on toe". This is obviously in allusion to the cobra's characteristic attitude when excited.

(b) English.—In India generally the snake is known as "the cobra" a word however applied in Portuguese to any snake. The Portuguese always referred to it as the "cobra de capello" or "hooded snake", and for many years subsequent to their arrival in India the qualifying adjective, now obsolete, was retained to distinguish it from the "cobra monil" or "necklace snake" (Russell's viper), the "cobra de aqua", or "water snake," etc., etc.

VARAD B. GIRI

(c) Indian.—In most parts of India the cobra is known as "nag", "nag samp," or some variation of "nag". Every juggler has a dozen qualifying terms, at his command, and no two jugglers will agree in the name they apply to a given individual. One hears of "arege nagoo", "coodum (wheat) nagoo", "jonna nagoo", based on colours resembling that of various cereals, "chinta" or "scinta (tamarind) nagoo", "malli (jasmine) nagoo, and "mogla nagoo" based on the names of plants, "cowri nagoo" and "sankoo nagoo" based on the names of shells, "kala nag," "sata nag", and a host of other "nags". Such names mostly emanate from professional snake men, and are of little or no interest, but there are many other local names that deserve mention. "Naya" is the name applied to the cobra by the Singhalese in Ceylon, but in this Island where so much Tamil labour is employed, South Indian names are frequently heard, such as "nalla pambu" (good snake), and "naga pambu". On the West Coast I heard "murukan" literally "cruel", and "sarpam", both Malayalam words applied to it. In Mysore the Canarese call it "nagara hava". In Bengal where the two forms, viz., the binocellate and the monocellate are associated, the former is known as "naga gokurra", and the latter as "keauthia", according to Fayrer, Ewart, Nicholson and others with "kala" or other qualifying prefixes. "Gorhmon" is another name used in Bengal for pale varieties. In Bihar "goh-manna" and "nag" are in use. In Rawalpindi "chajli" or "chajliwalla" is the common name for the cobra. "Chaj", I understand, is the Pushtu for a winnowing fan, which its dilated hood is somewhat fancifully held to resemble. Another Pushtu name in allusion to the hood is, I am told, common about Peshawar, viz., "chamcha-mar" (spoonsnake). The comparison to a flattened spoon is quite a good one, and has been noticed by people in other parts. "Mywe howk", the usual name for the cobra in Burma, implies "hissing snake". In the Chin Hills, the vernacular name is "tlua-kan". The cobra in Thailand is called "ngu how", "Toodong sta" is the name for it among the Malays at Kedah. Also "ular mata-ari" (sun snake), and "ular tedong sendok kunyit" meaning "turmeric ladle hood snake".

The typical cobra is yellowish to dark brown above, with black-and-white spectacle-mark on the hood and a black-and-white spot on each side of the lower surface of the hood. 25 to 35 scales across the neck, 23 to 25 across the middle of the body.

(a)   One or two dark brown cross-bands on the belly behind the hood.
(b)   Body variegated with darker, and lighter; belly with several dark cross-bands which may extend across the back.

In Shillong (Khasi Hills, Assam) I was told by three people independently, of a bright green cobra that was in the possession of a juggler shortly before my visit. Unfortunately the man had evidently left the station as I could not trace him. Curiously enough talking to Mr. W. Tottenham, Commissioner of Forests in Dibrugarh, a few days before this he mentioned a bright green cobra that he had encountered in Northern Thailand, at a place called Nan on the Mekong River, but which for want of spirit he could not preserve.

I have examined a curious variety of the cobra from the Andamans submitted to me by Dr. Annandale from the Indian Museum. It differs in coloration from anything I had previously seen, and does not conform to anything I have read of. It is fawn coloured and has well defined, broad, black, chevrons running down the back, the apices directed forwards, and the arms ending low in the flanks. The intervals are about one-third to one-fourth the width of the chevrons. Also in the flanks between each chevron is a small black triangular spot. The hood bears a light monocellus. The ventrals and subcaudals are 175 and 65, the scales in midbody 21, and the lepidosis as in a normal cobra except that the præocular shield fails to touch the internasal. If not already christened I suggest for this the name *sagittifera*.

Dimensions.—The cobra when adult measures usually from four and a half to five and a half feet. Larger specimens are rare, and six-footers extremely rare. Dr. Nicholson who for sometime distributed the rewards for poisonous snakes on behalf of the Mysore Government, says that out of 1,200 specimens that passed through his hands at Bangalore only 4 exceeded 5 feet 6 inches, and the largest of these measured 5 feet 8 inches. I have probably examined 500 cobras from various parts of Asia between Baluchistan and Chitral to South-China. I have only once seen a six-footer and this was killed in the Dun. It taped 6 feet 4 inches.

General characters.—The head is depressed, the snout rather short, with no canthus, and broadly rounded as seen from above. The nostril is rather large, and occupies the full depth of the suture between the nasal shields. The eye is moderate in size, and the round

shape of the pupil can usually be well seen in life. In some specimens however the usually distinct arc of gold at the papillary edge may be so faint that the pupil does not show up against the iris. There is a more or less obvious swelling in the temporal region over the underlying poison glands. The tongue is blackish at the tips.

The shields on the head are highly polished. A neck is scarcely if at all evident. Just behind the neck the hood commences, and is discussed later. The body is depressed, and there is a more or less distinct narrow groove down the spine. The scales over the back like those on the head are highly polished. The body maintains a fairly uniform girth throughout. The tail is short, and accounts for from one-fifth to one-ninth the total length.

Identification.—This is easy if attention is paid to the lepidosis (=scales). The presence of the little "cuneate" scale between the 4th and 5th infralabials will declare the snake a cobra among all *land* snakes. It is found in many of the sea snakes however. Rarely two such scales exist in the cobra, but it is very rarely absent on both sides.

If a typical hood mark is present either of the monocellate or binocellate type the diagnosis is also easy, but in many specimens there is no hood mark at all, and in death rigor mortis stiffens the joints so that the hood is not easily demonstrable. It is for this reason that attention to shield characters is recommended. If the hood is seen dilated in life there should not be any doubt about the snake, but it must be remembered that the hamadryad *Ophiophagus hannah* has a well developed hood and that other snakes flatten the neck to a more limited degree. The cobra has been confused with the *Pseudoxenodon macrops* of the Eastern Himalayas and Assam, and with the *Argyrogenia fasciolata*, a snake common in the Western Ghats, but also known from the Eastern Ghats, and the Ganges Valley.

Disposition.—The cobra is usually not an aggressive snake. When flushed in its native haunts it nearly always tries to escape, and usually succeeds in doing so, but is often shot before it gets to a place of safety. I have encountered many, and find that at close quarters if suddenly disturbed, or it may be if stepped upon, it quickly erects itself, hisses loudly, sways backwards and forwards and awaits its opportunity to strike. If one keeps still, the menace is quickly over, and the snake drops its head, and slinks off. An incautious movement however causes it to turn, erect itself once more, and challenge the intruder again. Many good observers have remarked on its timid nature.

Spectacled or Binocellate Cobra
Naja naja

Black Cobra
Naja oxiana

Monocellate Cobra
Naja kaouthia

The cobra is sometimes very fierce, and when disturbed may be a very dangerous snake to encounter. Whatever spirit and aggressiveness may be natural to it in the early days of captivity, I think all will agree that it is very easily tamed. This is evident to anyone who has seen jugglers, and professional snake-men with their captive specimens. If a specimen has been on show for long, it will often require a good slap on its back to provoke it to erect itself and hiss. The cobra that will do so without such treatment one may depend upon it has been but recently deprived of its liberty.

Young cobras are much more dangerous than adults as a rule. They seem more on the alert, more easily excited, and strike repeatedly and with much malice.

It is certainly significant that one never sees a young cobra in the hands of jugglers.

The cobra's effective striking range is a very limited one. I believe the erection of its forebody and the expansion of its hood are invariable preliminaries, and the height to which it can erect itself forms the radius of its stroke. This radius when the snake is erect is very deceptive, appearing much greater than it proves to be when measured along the ground on the completion of its stroke. Jugglers from long practice estimate this range wonderfully, and contrive to evade their captive menace, with remarkable precision, withdrawing their hands often only a few inches from the spot where the stroke is delivered.

The bite is often a mere snap of the jaws, and the bitten part immediately released, but sometimes the snake will fasten itself tenaciously, necessitating a forcible opening of the jaws to effect release. I was told of an incident where a young sampwalla was bitten, and the snake hung on to him so that it had to be removed by forcibly prizing open the jaws. Sometimes after a bite a drop or more of venom may be seen on the skin of the bitten subject which may be wiped off without gaining access to the punctures inflicted. More rarely poison is shaken off in the form of a spray or jet by the forcible thrust forward of the snake, which may fail to reach the object of its attack.

I have on more than one occasion witnessed this with jugglers who unconcernedly wiped away the poison emitted. A Hospital Assistant of mine, whilst trying to dislodge a cobra that had taken refuge in the wall of his garden, had a jet of poison ejected into his face. One may presume that some such incident caused Boie to christen one variety of our Indian cobra *sputatrix* (spitter). The habit is well known among certain African cobras, I believe the venom ejected is shaken off the fangs, and carried forward by the vehemence of the thrust. In some instances, however, where a shower of spray is reported it is more probably caused by the explosive expiratory blasts from the glottis which occur while the snake is hissing, and to which I refer again later.

Haunts.—The Cobra may be found almost anywhere. I have encountered it in heavy jungle, and in open country far removed from forest growth. The ryot disturbs it in his crops, the mali in Cantonment gardens, and the sportsman when quail, partridge or hare shooting. It is a common snake in almost every populated area, and I have had it sent to me frequently from within Cantonment limits, from the regimental and other bazaars, from Artillery and other lines, the suburbs and actually in the gardens of our largest towns, from inside jails, the godowns of the Supply and Transport Corps, and the Telegraph and such like departments, from the warehouses of various mills, and such like situations. No amount of bustle or disturbance seems to deter it from taking up its abode in man's immediate vicinity. It was sent in to me several times in Rangoon from timber yards, where hundreds of men were working daily, elephants pounding up and down moving timber, engines vibrating and throbbing and circular saws screeching through boles of teak. Even in such scenes of turmoil it will establish itself beneath a stack of wood, or convenient drain, and escape dislodgement for long periods.

Old masonry invariably harbours cobras among other snakes. In Delhi the old walls of the Fort were always a safe draw for the snakeman whom I saw every week bring in his bag—some half dozen or more—to be robbed of their poison, which was being collected for the Government of India. Similarly old cemeteries, and ruined habitations, mosques, etc., furnish ideal quarters for this snake. Another favourite haunt is the loose brick work of old wells. The basements of many houses in Cantonments and bazaars can boast a cobra tenant, and it is not surprising therefore that this snake is so frequently encountered inside bath-rooms, and dwelling rooms, besides stables and servants' habitations. Further afield an ant's nest is often a specially favoured resort, or it may be any hole in the ground, or at the base of a tree among its roots. It is frequently found near water, and often actually in that element, in which it swims with facility and strength. A few cases of cobras climbing trees are reported, the object usually being the plunder of some bird's nest.

It has been occasionally reported in the sea, perhaps carried thence by rivers in flood time, but sometimes no river being in the vicinity it must have taken to the sea of its own free will. In one instance a four-footer is said to have managed to board a man-of-war, viz., the *Wellington*, lying off the Coast of Ceylon at Aripo. Another account of the incident however, says that the sailors saw the snake in the sea swimming vigorously towards the ship, and assailed it so successfully with billets of wood and other missiles that it returned to land.

Food.—The cobra feeds principally on rats, frogs, toads, and less frequently on birds and it seems to show no special preference for any of these creatures. Its choice in batrachians is largely determined by their size, the most bulky individuals being apparently those most sought after. Thus among frogs it is the bull frog, *Rana tigrina*, which is most usually victimized, and among toads *Bufo melanostictus* and *B. andersoni* receive special attention, and internal accommodation. Rats and mice are very frequently taken, and I think there can be no doubt the numbers of these vermin are materially checked by this snake. I was astonished in Bangalore some years back to see with what avidity the captive cobra belonging to a juggler accepted dead mice which he withdrew from his pocket. The man offered them as one would a morsel to a dog and one of his cobras nosed its snout into his hand, and took three mice, swallowing them one after another in a couple of minutes or so. Other creatures are taken as circumstances dictate. The Rev. C. Leigh, s.j., writing from Trichinopoly of his captive specimens told me that after eating two small frogs, and then three middle sized ones, one cobra finally disposed of two squirrels (*Funambulus palmarum*). Sometimes birds are attacked, and killed, especially poultry. In Fyzabad one got under a hen coop one night in a native hut, and killed the hen and six chicks. The snake met its death the next night, swallowing a frog baited on a hook. On another occasion one got into a quailery in Fyzabad, and accounted for 13 birds in the night. One only of these had been swallowed, and it seems to me likely that some or all the rest may have died from fright. Only recently in Almora an officer whilst quail shooting flushed a cobra which he shot in attempting to escape down a hole. The snake was cut in half by the shot and a freshly swallowed quail fell out of the stomach. One caught at Trivandrum was enormously distended and contained a monitor lizard (*Varanus bengalensis*) two feet long. Occasionally the cobra exhibits ophiophagus tastes. One in captivity ate another with which it was caged, both snakes having seized the same frog, and commenced eating from opposite ends. On another occasion one was observed to eat a wolf-snake (*Lycodon aulicus*). Here I may mention that the cobra itself sometimes falls a victim to its larger and more confirmed ophiophagus relative the king cobra *Ophiophagus hannah*.

Some interesting accounts have appeared of cobras eating the eggs of poultry. The egg of a guinea fowl recovered from a cobra's interior was set, and in due course hatched out.

A cobra that had got into a guinea fowl's nest, had eaten 6 of the 15 eggs. The eggs were subsequently removed, and set and 3 eventually hatched out. After the publishing of these events Colonel Bannerman experimented on cobras in the Parel Laboratory,* to ascertain how long it took for the egg shell to dissolve under the influence of the gastric juices, and found that it required about 48 hours. Inspection of the subsequent excrements showed in one case that a few pieces of egg shell were discharged 16 days after the experiment.

In captivity many specimens feed eagerly, and thrive well. On the other hand it would appear from the methods of some professional snake-men that they find some of their specimens difficult to tempt with food, for many carry with them a small natural funnel which appears to be part of the shaft of the tibia of a goat, which they insert into some cobras' throats, and into which they break a fowl's egg, or pour milk.

Habits.—The most notable habit of the cobra is the very remarkable pose it adopts when alarmed and which has gained for it world-wide renown. Not only does it erect the forebody to a remarkable degree, but it flattens its neck in a very remarkable, and characteristic way to form the so-called "hood". The height to which a cobra can erect itself is usually very much overestimated by the casual observer. I have taken careful measurements on several occasions, marking off the height on a stick when the snake's attention was engaged by a juggler. The measurement of the whole snake in life is not easy, and the lengths given must not be taken as very exact. I found the degree of erection commensurate with the degree of excitement or provocation. One snake measuring 5 feet 2½ inches poised vertically to a height of 13 inches, another 5 feet 4 inches long sat up 15½ inches, a third 6 feet and ½ an inch raised itself 14 inches, and a fourth 5 feet 1 inch, only 7½ inches. On the 20th August 1904 in Bangalore I found a 5-footer just sat up 15 inches, but on the next day in the presence of a mongoose that was causing him much agitation the same snake erected itself 21 inches. Another cobra 1½ inches less in length raised itself just 21 inches under similar provocation. It may be taken then that the maximum limit of erection is about one-third the length of the snake.

The so-called "hood" is formed by the action of muscles operating upon the ribs in the region behind the neck. I have examined a skeleton in the museum of the Royal College of Surgeons, London, which is well set up in the erect position, and with the ribs fixed as they would be in the expanded hood in life. The atlas (1st vertebra), axis (2nd vertebra), and the 3rd vertebra have no ribs, but the 3rd has an elongate rib-like transverse process. The succeeding 27 vertebrae have ribs attached to them that are involved in the production of the hood. These ribs are much less bowed than those in the rest of the body, and enjoy a range of movement greatly in excess of the other corporeal ribs. The 9th is the longest on the left side, and measures 41 mm., and the 10th measuring 42 mm. is the longest on the right side. The preceding and succeeding ribs progressively diminish so that an oval outline is given to the hood. The ribs are set obliquely forming an angle of 40° to 45° with the long axis of the spine. In the prone state they are directed backwards, outwards, and downwards, and give a contour to the body almost like that in other parts. In the erect pose the corresponding direction of the ribs would be downwards, outwards, and forwards, but any forward tendency is entirely obliterated by the action of a set of dorsal muscles that not only draw the ribs back till they are completely transverse,

---

* Haffekine Institute, Parel, Mumbai

but also fully straighten them. During full expansion, judging from freshly dissected hood, I think the ribs are also slightly elevated, and the angle made with the spine thereby rather increased. As the overlying skin is but loosely attached it does not in any way hamper the movements of the ribs within, which by their backward extension and elevation enormously stretch it in a lateral direction, at the expense of the ventro-vertebral diameter. The oval shape of the hood, and the flattening produced has been well compared to a shallow spoon, or skimmer. The hood originates high up in the nape, and the head bent strongly at the atlas joint is carried at right angles to it when spread. The arching of the forebody and general pose and movement of the cobra when erect remind one very forcibly of the carriage of a swan's neck. The dorsal skin is very much stretched when the hood is expanded, so that the scale rows are widely separated and as the hood marks are almost entirely confined to the skin they become very conspicuous. The curious pose adopted can be sustained for a considerable time, certainly many minutes if sufficient stimulus is offered, and continued. Whilst poised with expanded hood the snake sways restlessly forwards and backwards and can be made to bend backwards to an extraordinary degree before losing its equilibrium. It hisses in a fierce explosive manner whilst erect. I have carefully observed caged specimens at this time. I noticed that hissing occurs both during inspiration, and expiration. The inspiratory is the shorter act, and its note higher pitched than the expiratory. It is quavering in quality, reminding one of a knife on a grindstone. The expiratory effort is the longer, louder and lower pitched and intermittently explosive in character. The tongue is flicked out during both inspiration and expiration. The inflation extends as far forwards as the chin shields. Whilst erect the snake inflates its body independently of its hood action, and the inflation affects nearly the whole body length, declining posteriorly till finally lost a few inches before the vent in an adult.

One of the most interesting matters in connection with the cobra affects that ever fruitful subject of discussion, "charming". It is clear that many very competent authorities disbelieve in the practice. Many good observers state that it is the constant movement of the musical instrument in front of the snake that keeps it erect and not the noise produced. I certainly take this view myself, and came to this conclusion very early in my Indian career. One thing puzzled me at first and aroused my suspicion, viz., why is it that in all the stories one reads of "charming", it is invariably the cobra that withdraws from its snug retreat, whilst other snakes apparently are not susceptible to the captivating(?) sounds of the juggler's pipe? I know of no anatomical difference in the auditory apparatus of cobras from other snakes. I experimented frequently in Delhi in my verandah with cobras. I cut narrow strips of sticking plaster, sufficiently broad to cover the eyes completely. These strips had a double purpose. Not only did they blindfold the subjects of experiment, but being carried right round the head they locked the snakes' jaws, and so prevented any chance of my being bitten. This done the snake was released, and in a very short time it relaxed its hood, and assumed a completely recumbent attitude. The verandah in which the first of these experiments was carried out was a crazy wooden structure, and if one moved a chair, or even if a servant walked along the room inside, the snake immediately erected itself as if conscious of danger. On the cement verandah downstairs, it was also noticed that the snakes immediately got up when any one walked along in the near neighbourhood. I had a kerosene oil tin at hand, and when the snakes were recumbent I beat this with a stick close to their heads without their taking any notice whatever. Similarly I blew a bugle close beside them, and if an amateur's attempt at bugling failed to rouse them they must indeed be deaf. The greatest care is necessary in conducting such experiments, to eliminate all other possible means of rousing the snake. For instance if a

rusty tin is beaten over the snake, particles will fall on it, and rouse it to attention. Similarly if the blast of air emitted from the bugle impinges ever so little on the snake, it is roused to action, and erects itself.

Many people suppose that a snake is deaf, but this is not the case. Snakes hear well though they have no external ears. Many people are not aware that there are two ways in which the essential auditory apparatus may be stimulated, and sounds heard. If one strikes a tuning fork, and places the stem on any part of the skull, or even the spine to its lowest part, the vibrations can be heard distinctly. If the head is in contact with a table, and the tuning fork struck, the sound is audible when the stem is placed on the table at some distance though inaudible when not touching the table. This is due to the conduction of vibrations through solids and such vibrations are better heard, and for a longer time than those conducted by waves of air which strike upon a membrane "the drum", situated at varying depths (according to the particular animal) in a canal in the skull (the external auditory meatus). The drum set vibrating acts through a chain of tiny bones in the middle ear, so as to affect fluid contained in semicircular canals in the internal ear, the fluid in its turn communicating the vibrations to highly specialized sense organs at the termination of the filaments of the auditory nerves. These nerves carry the impulses received to the brain centres where they are interpreted as sounds. This latter method of conduction, viz., by means of the air is the predominating one in mammals, birds, and many reptiles, but is entirely wanting in all snakes, there being no external orifice, and no drum to receive impressions. Conduction by solids is however good in snakes, perhaps for all we know more highly sensitive than in man.

Now it is obvious that if snakes have no ear openings and no drums they cannot hear sounds conducted by air, such as those emitted by musical and other instruments. This accounts for the cobras taking no notice of the noises I made at close quarters, though they were keenly alive to sounds such as footsteps communicated through the ground. If one is to believe the wonderful stories, told in good faith I have no doubt, about "charming", one must explain it by assuming that snake charmers are possessed of some occult force not apparent to the spectators, for it cannot be explained through the agency of sound conducted by air. As a matter of fact a snake charmer in Bangalore with whom I had become very familiar admitted to me that snake-men knew that snakes were

Albino Indian Cobra

M. KRISHNAN

deaf, and that the whole of their "charming" was a hoax. It is most certainly the incessant movement of the man's arms while piping, or the restless movements of his knees while squatting that affords the necessary stimulus, and keeps the cobra excited, and erect.

It is very curious how all absorbing movement is to the cobra, you have only to attract its attention with one hand, while you seize it in the middle of the body with the other and the snake is yours. It strikes *in every direction especially at any moving object*, but it never seems to occur to it to turn, and bite the hand that is holding it as almost all other snakes would do at once. I consider this strange trait argues a very great lack of intelligence.

The cobra seems to show a decided tendency to a social life. Many writers have remarked upon its habit of living in couples, and this is specially true during the breeding season. It appears however to seek society apart from sexual impulses for on one occasion in Rangoon two were brought to me found coiled together beneath a stack of wood, and both proved to be females. On another occasion, also in Rangoon, a Burman dug out a hole where he had seen a snake make good its escape. The result was the discovery of three cobras. Two of these were males, and one a female which showed ovarian follicles obviously fertilized and enlarged. This leads one to ask the question does the cobra on occasion practice polyandry?

The cobra is frequently abroad during the day. I have several times met one when bird nesting, shooting or out after butterflies. Many of these were obviously not roused from a siesta, but were roaming about I suppose in search of food or drink. In populated areas it is perhaps more frequently encountered at night.

Like other snakes it suffers from thirst, specially in the hot weather, and I daresay that many of its intrusions into bath-rooms and its lodgement in catch-pits and wells may be accounted for in this manner. I saw one in the possession of a snakeman in Cannanore that dipped its head into a tin of water presented to it, and drank greedily, each gulp being plainly visible in the throat.

The Sexes.—I can discover no difference in the lepidosis (scalation) of the sexes, nor in the relative lengths of the tails. There is no constant difference either in the ranges of ventrals or subcaudals. The male clasper is narrow and long and surmounted with very small claw-like tentacles. It is not bifid. Females appear to be more numerous than males in Bangalore, as Dr. Nicholson found 410 of the former against 308 of the latter in 718 cobras sexed by him.

Breeding.—The mating season extends over several months of the year. Flower in Siam had a gravid female with eggs fit for discharge judging from their measurements in the month of January. Nicholson had several gravid females with eggs about an inch long in February at Bangalore, and I had one in a similar condition at Cannanore in the same month. Mr. E.E. Green also had a gravid female in Ceylon in the same month. Colonel Dawson had captive cobras, in Trivandrum, which were observed "in copula" in January. Mr. H. Hampton wrote to me of a pair he had in captivity at Mogok, Ruby Mines, Burma that were observed coupled at the end of March. Evans and I obtained gravid females in Rangoon in July and August, one specimen in July showing but little enlargement of the ovarian follicles. Mr. Foulkes told me some years ago of a pair reported coupled in June at Rajamundry.

The act of mating has been witnessed by Colonel Dawson and Mr. H. Hampton to whom I am indebted for the following details. In Trivandrum the pair remained coupled from 11 a.m. and coitus lasted intermittently for three days. He observed that the pair nodded their heads continually, and their bodies quivered. They did not take the slightest notice of anybody in front of their cage. They did not expand their hoods, neither did they wrap themselves around one another. Each turned the vent upwards and sideways to effect engagement.

Period of gestation.—The cobra is known to be oviparous, and the period of gestation is accurately known in Colonel Dawson's case. Sixty-two days after coitus, i.e., on the 20th of March, eight eggs were deposited, the first at 8 a.m., six more immediately, and then after the lapse of half an hour the last. In Mr. Hampton's case the mating was observed towards the end of March, and eggs were not deposited until the middle of August, nearly 5 months.

Spectacled Cobra

Season of Egg-laying.—The usual month for the deposition of eggs is May. All the eight cobras that have laid eggs at Parel Laboratory did so in that month. Nicholson, too, says that about Bangalore they are laid in May and early June. Mr. Phipson remarked that eggs are laid in the rains, and Fayrer, too, says that his snakemen told him that about Bengal they laid eggs in the rains. Two eggs sent to me from our Society's collection were deposited in June. Wall (A.J.) mentions eggs laid in July, and Hampton's eggs were laid in the middle of August at Mogok.

Size of Eggs.—The eggs are elongate, white, ovals with soft shells, and similar poles. The two sent to me from our Society measure 49 x 28 mm. (a shade under 2 inches in length). The almost mature eggs extracted from the maternal abdomen by Flower measured 53 x 34 mm. Eggs sent to me from Parel vary much, and are much smaller, and it occurs to me they may have been infertile. Two of these measured 41 x 20 mm., one 38 x 19, one 32 x 20, and a fifth 29 x 15 mm.

Number of Eggs.—From over a dozen records, I find that the usual number of eggs laid is 12 to 22. I find one record of 8, and the only record of over 22 is Mr. Hampton's. In this case 45 eggs were deposited, 36 seemingly good, and 9 apparently infertile.

Incubation.—Fayrer says on the evidence of his snakemen that the cobra incubates her eggs, and that they frequently dug out mother and brood. This is in accordance with the habits of other snakes and receives direct confirmation from Colonel Dawson, who told me that at first his dam coiled herself among her eggs. The period of incubation has been ascertained at Parel. Eggs laid on the 12th May hatched out on the 20th of July, i.e., in 69 days. The period that elapses then between coition and the advent of the young is rather over 4 months.

Hatchlings.—Mr. Phipson reported young measuring only $7\frac{1}{2}$ inches as they emerged from the eggs in our Society's rooms. All the other testimony at my command agrees in assigning to the hatchling a length of 10 to 11 inches. Assistant Surgeon Robertson told me the young he saw just hatched measured 11 inches. I measured one of those that hatched at Parel, which was bottled at once, and found it was $10\frac{1}{2}$ inches long. Nicholson remarks that at birth they are less than one foot. Now Colonel Bannerman extracted an embryo from an egg 43 days after deposition, and found it taped 7 inches. Another that was removed from an egg by me measured 9 inches; but it is not specified at what lapse of time after deposition. It is curious from these two last specimens to account for Phipson's hatchlings only measuring $7\frac{1}{2}$ inches. I have had young cobras brought to me measuring $10\frac{1}{2}$ inches in June at Cannanore and $11\text{-}\frac{1}{8}$, $12\frac{1}{2}$ and $12\frac{3}{4}$ inches at Fyzabad in July. Nicholson remarked that out of 1,000 cobras brought to him in May to August 1873, 230 were young of the season measuring from 12 to 16 inches, and of 1,220 in the year 50 were from eggs deposited. It seems to be a common belief that young cobras newly hatched are not poisonous. This is certainly a mistake, as Mr. Phipson reported that the young cobras that hatched out some years ago in our Society's rooms killed a small Malay python (*Python reticulatus*) which was placed in their cage, a few days after they were born. They attacked it at once, biting it viciously across the back.

Growth.—Phipson referring to the hatchlings that were $7\frac{1}{2}$ inches when they emerged from the egg, says they grew an inch and a half in about two months, but as these specimens appeared to have died of hunger, having refused all food, one may be certain this underestimates the normal growth. Similarly, I have had specimens submitted to me from Parel which did not develop as cobras usually do in a state of nature. Four of these born on the 18th July 1910 were consigned to spirit on the 2nd November. I measured these, and found them $11\text{-}\frac{9}{16}$, 12, 12, and $12\text{-}\frac{5}{8}$ inches. A fifth specimen born on the 20th July 1910 died on the 7th December, and I find it is $12\frac{3}{4}$ inches long.

Nicholson's observation shows that young measuring less than a foot at birth attain a length of from 2½ to 3 feet by the end of their first year of life. This rate of growth is out of all proportion to that noted by me in connection with other snakes, and I expected to find some error in his conclusions. My own notes, however, confirm Nicholson's statements. I find that young averaging 12 inches in July, average 2 feet 6 inches by the next July. At the end, of their second year they average 3 feet 8 inches, at the end of the third 4 feet 2 inches, and at the end of the fourth 4 feet 10 inches. The growth, it will be seen, is especially rapid during the first year, and progressively diminished in subsequent years. In other snakes I find it the rule that the young proximately double their length in the first year.

Sloughing.—Fayrer mentions a cobra that cast its skin on October 17th, and again on November the 10th, and December the 7th. Another in his possession desquamated on the 15th of October and on the 6th of November. In Trivandrum a captive cobra shed its skin on November 10th, 1902, and on February 19th, April 8th and July 28th in 1903. I have been told by snake-men that ecdysis occurs about once a month, and Vincent Richards gives about the same period between successive moults from his observations. It will be seen from the above that there is no regularity in this function, which may occur at intervals ranging between three weeks, and three months. I am informed by snake-men that specimens in captivity sicken during this period, and that they are afraid to give them food or drink as it upsets them. They certainly appear very dull, and non-captive specimens are most likely to meet with their death, if they venture out of their holes at this time, the disc before the eye becoming so opaque that the creature is virtually blind for some time.

Foes.—Among mammals, the Mongoose has been conceded a special place as a destroyer of cobras. Personally I always had the greatest difficulty to get my captive Mongooses, and I have had three or four, to face my captive cobras, much less attack

Source: Indian Serpents, by Patrick Russell, Publisher East India Company, 1801.
The first book on Indian Snakes

them. Mr. Stevens in Assam told me he once witnessed an encounter between a Mongoose and a cobra. The snake managed to evade the carnivore in the tall grass and was killed by Mr. Stevens. An interesting incident was reported to me by Mr. Reid showing that some animals have an instinctive dread of the cobra, or perhaps snakes in general. A herd of buffaloes that were standing, feeding out of a row of "nands", suddenly became very excited, and broke loose, stamping and snorting, and to all appearances were terrified. On investigation a cobra was found close by which was killed, one old cow when she saw it rushed upon its body, and trampled it. This by the way is the method by which deer and pigs are reported to attack and destroy snakes. Gunther says the jungle fowl kills young cobras, and this seems probable, as domestic fowls are known to kill and eat them; an event of this kind happened before the eyes of the late Mr. P.W. MacKinnon in Mussoorie, his fowl killing, and then swallowing the snake with no ill-effects. Both Evans and Craddock have reported instances of the cobra being victimized by the king cobra *Ophiophagus hannah*. Mr. Gleadow once wrote to me that he saw a large monitor lizard (*Varanus* sp.) running off with a live snake, 3 or 4 feet long, in his jaws, which when released was shot, and proved to be a cobra.

The fangs.—In Chamber's Encylopaedia the article on the cobra says that its fang is not canaliculated, but grooved. Mr. Boulenger too in his catalogue refers to the fangs as being grooved and they are shown with a deficiency in the anterior wall in the figures in Fayrer's and other works. This is most certainly not correct. The fact that there is an indistinct line on the anterior face of the fang does not affect the question of its being canaliculated. The line referred to is a seam which marks the spot where the circumflexed walls of the canals meet and blend. There is a considerable opening at the base of this seam and a much smaller one near its point, where the poison finds exit. It is not generally known that it was this beautifully specialized instrument in the jaws of a poisonous snake, that led a medical man to design the surgical instrument used so freely in these days in the form of the hypodermic needle.

The cobra's fang is relatively small compared with viperine fangs and is a much more solid and stronger weapon. The length of my largest cobra fang is 7 mm. and was taken from a large adult. The length of the fangs in a fifteen inch *Echis* (Phoorsa) in my collection is 5 mm., and those in a 3 feet 4 inch viper (*Lachesis anamallensis*) are 13 mm. My largest hamadryad measuring 11 feet 5 inches had fangs 10 mm. in length.

There are usually two fully operative fangs fixed in each maxilla, but these are shed singly at intervals, and from Fayrer's experiments 18 days was the shortest period that elapsed between drawing them, and the fixation of a new one.

The poison gland.—This organ, which is really a salivary gland, and the analogue of the parotid gland in mammals including man, consists of a body and a neck. The body is much the shape and size of an almond, and consists of (1) a thick fibrous capsule or jacket, (2) the glandular or poison secreting substance proper, and (3) a duct running centrally in the long axis of the gland. The capsule gives off numerous fibrous septa which pass into the glandular substance and divide the gland into numerous chambers or pockets (the poison lakes of Bobeau). Each pocket is lined with poison-secreting cells, and carries in its walls blood vessels which convey the blood upon which the poison cells depend for their activity. After a period of activity, the pockets which converge forwards and inwards towards the axis of the gland, become distended with poison, and this is poured into the central duct. At the posterior pole the gland ends in a downward projecting lobe. The fibrous capsule dips into the gland just in front of this lobe to form a furrow for the attachment of an important muscle, the masseter. This muscle originates from the postfrontal bone, and the ridges on the parietal, and is somewhat fan-shaped. Its fibres

converge, and pass first backwards over the superior and internal surfaces of the gland, then downwards behind its posterior pole, and finally forwards to be attached to the furrow, or dimple in front of the lobe. The muscle, in fact, embraces a large part of the gland surface, and in contraction squeezes it much in the same way as the hand operates on a bicycle horn, the result being that poison is driven forwards into the duct to pour finally into the mouth. The neck of the gland consists of a sheath which is the direct continuation of the capsule surrounding the body, but is much thinner. Centrally is the poison duct, and intermediate between the sheath and the duct a series of mucous glands. These are placed at right angles to the axis of the duct, and discharge their mucous into that channel where it mingles with the poison proper. In section the gland appears to the naked eye much like a sponge.

Physical characters of cobra poison.—Cobra venom when freshly secreted, is a clear, amber-coloured, very viscid fluid with a specific gravity of 1,050. It resembles olive oil in appearance and consistency and soon solidifies into an amorphous, brittle mass, fissuring in all directions, and losing from 60 to 75 per cent of its weight in the process of drying.

A drop of fresh cobra venom weighs approximately 35 mgms. Allowing a loss of 68 per cent in drying the residue of one drop would weigh 11 mgms. It is somewhat remarkable that a drop of olive oil to which I have compared cobra venom in appearance and consistency only weighs 6 mgms. in the same balance at Parel. When dry the poison retains its transparency and resembles gum or amber. In the dry state it keeps well, and preserves its virulent character according to Vincent Richards for at least 15 years. The same authority shows that though there has been some difference of opinion among authorities, poison is acid when fresh, and this, in spite of the fact that the normal reaction of the cobra's mouth is alkaline. Lamb has confirmed these observations. After the lapse of some hours the venom becomes neutral. Dr. Nicholson says it is slightly bitter to the taste and causes a feeling of frothy soapiness in the mouth, at the same time stimulating the flow of saliva. The same authority remarks that the dried particles have a pungent action upon the nostrils. Lamb describes the taste as very bitter and astringent.

Quantity of poison secreted.—This, of course, varies with the size of the cobra, but even in specimens of similar length other factors affect the yield. Lamb says: "it is an observation of common occurrence in this laboratory (Parel), that a cobra newly caught will yield from 20 to 30 large drops of poison, while after he has been a captive for some time this quantity will have diminished to from 6 to 10 drops and in time to nil." Dr. Nicholson observed that the yield was more abundant in wet weather. Under the influence of anger poison is secreted unduly copiously. Doubtless age, health, and individual vitality also influence the quantity secreted.

The venom in its fluid state is found to vary a good deal in concentration, a cobra's yield is therefore calculated by the amount of solid residue left after drying. Cunningham's average for 9 cobras was 254 mgms., Lamb's for 14 cobras 231, and Rogers's for 2 cobras 249 mgms. Lamb found that by provoking cobras, so as to make them bite viciously, the yield collected in glasses was considerably augmented, as compared with that collected by simple pressure over the glands, and amounted to an average of 373 mgms. for 3 cobras. Cunningham obtained from one cobra the enormous quantity of 726 mgms. The amount of solid, it will be seen, ranges between 200 and 726 mgms. in healthy adult cobras.

Toxicity of cobra poison in man.—Lamb has shown that even the dried product varies in its degree of toxicity, as he found the minimal lethal dose for rats was 7 mgms. with one sample and 4 with another.

Lethal dose of cobra poison in man.—It is a well established fact that cobra venom may be swallowed in large quantities without producing any baneful results. Elliot gave a dog 10 drops—a dose sufficient to kill 10 dogs, if injected into the tissues—without producing any ill-effect. On another occasion he gave 20 drops to a goat with the same result. Fraser by graduated doses internally succeeded in giving a cat 1,000 times the lethal dose by injection beneath the skin. Calmette repeated the experiment, giving 1,000 times the lethal subcutaneous dose to a cat internally, without producing symptoms of poisoning. It is the access of the venom into the bloodstream that constitutes its extreme danger to all animals. The lethal dose of the poison so introduced has been accurately ascertained for many animals by experiment, but in man must remain to some extent conjectural. Various estimates have been made, based on experiments on the lower animals. Fraser's estimate is 31 mgms. whilst Calmette made it about 10 mgms. Lamb however, finding that 25 mgms. is proximately the minimal lethal dose per kilogram weight in monkeys, and postulating an equal degree of susceptibility in man, concludes that the dose for a man weighing 10 stone would be about 15 to 17.5 mgms. If we take Lamb's estimate of the lethal dose for man, which is probably nearest the mark and strike an average for the 25 cobras experimented with by Cunningham, Lamb and Rogers, the average yield of which amounts to 240 mgms., we may state that an average cobra contains poison enough in its glands to kill 15 men. An exceptional cobra may even contain sufficient poison to kill 45 men.

Rapidity of absorption of cobra poison.— Blake found that a poison injected into the jugular vein, reached the pulmonary circulation of a dog in from 4 to 6 seconds, and the cardiac circulation in 7 seconds. A poison injected into the same vein was distributed throughout the circulation in 9 seconds. It is this extreme rapidity of transmission in the bloodstream that accounts for the fatal issue in experiments where a poisoned member is amputated or wound excised almost immediately after being bitten.

There is abundant evidence to show that snakes like the cobra, which are known to be capable of delivering a mortal wound, frequently fail to do so, though they may inject poison in considerable quantity. So far as the human subject is concerned there are many cases of cobra bite recorded, where no ill-effects were produced, or symptoms of varying severity, not ending in death, though no treatment was attempted. Dr. Davy after remarking that the effects of cobra bite "vary a good deal according to circumstances not easy to calculate", says: "I have seen several men who have recovered from the bite of the hooded snake, and I have heard of two or three only to whom it has proved fatal." Russell mentions a woman who he saw 10 hours after being bitten by a cobra. She recovered completely. He mentions another cases of a drunken Irishman who declared he was proof against any snake owing to his nationality, and put a cobra into his shirt before an assembled throng. The snake bit him severely in the breast, and he suffered not only great pain locally, but serious constitutional effects, nevertheless he recovered. Dr. Nicholson records a case where two snake-men under the influence of drink got bitten by one of their cobras. As some time had elapsed when he saw them he coloured some water pink with his dentifrice which he gave them to allay their fears. Both recovered, though one had a swollen hand next day as a result of the accident. Calmette records another very interesting case where a man was profoundly under the influence of cobra poisoning following a bite but who persistently refused antivenine which was at hand, took his chance, and recovered completely.

These cases are most instructive, and serve to point two lessons. One is that however serious the symptoms arising from a cobra bite, there is always hope. The other lesson is that nobody is qualified to assume that any given treatment adopted in a certain case has been responsible for its favourable issue. There can be no doubt that the failure to realize

this latter truth has been responsible for the host of reputed antidotes, which have been vaunted from time to time since the days of Celsus, all of which have proved futile when subjected to scientific experiment. It is difficult to say in what percentage of cases of Cobra bite would not prove mortal. Dr. Davy, speaking of Ceylon cobras, says that recovery follows the Allbutt's system of medicine, Lamb and Martin say "the mortality in persons bitten by the larger snakes of India and South America would not, from the scanty records available, appear to be more than 30 per cent."

Symptoms of cobra toxaemia.—These may be divided into local and constitutional.

Local.—The first, and perhaps invariable symptom, is pain, which is of a stinging or burning character, out of all proportion to the mechanical injuries sustained. It comes on immediately and persists, perhaps lasting for hours. If pain is experienced only to the degree excited by ordinary pricks or scratches, and is but transient, there is a jusitifiable presumption that poison has not been introduced. Coincident with the pain, and almost as speedy in its appearance is swelling which gradually increases until perhaps the whole limb is puffy. The third invariable sign that venom has gained access to the wounds is the oozing of a blood-stained serum. If on the other hand the punctures are sealed with clot as in ordinary wounds shortly after injury, there is every probability, if not actual certainty, that poison has failed to find entry into the tissues. The fourth cardinal sign is one which cannot be detected until the tissues in the site of the wound have been cut into, though it may be inferred if rapidly ensuing swelling has occurred, accompanied with the other signs. The tissues assume a very characteristic appearance, the parts become purplish centrally, the colour fading to scarlet, and then pinkish, and a thin serum exudes. In one case, Wall (A.J.) found this purplish effusion, which is characteristic of the action of snake venom, was seen within 30 seconds of the injection of the poison. When present, it is absolute proof of the absorption of venom; if absent, it is probably equally good proof of the failure of the poison to have reached the tissues.

How intensely irritant the venom is locally is apparent from the rapidity of the symptoms noted above, added to which is the fact that in many cases where the bitten subject recovers, the tissues involved actually mortify, and are thrown off as a slough. Occasionally one sees people with withered limbs stated to be due to the effects of a snake bite.

The constitutional effects are a gradual, but rapidly advancing paralysis, due to the action of the poison on the brain and cord. Sooner or later the bitten subject complains of weakness in the legs, and is prompted to recline rather than walk or sit. This weakness creeps up the trunk, and affects the muscles of the neck, so that the head droops, the muscles of the tongue, lips and throat, so that speech becomes difficult, the lips fall away from the teeth, and allow the saliva to dribble, and swallowing becomes difficult or impossible. The eyelids too droop giving a sleepy expression to the face. While these paralyses are waxing, the respiratory function becomes affected, breathing becomes difficult, then laborious and finally death from respiratory failure ends the scene. Among other toxic symptoms may be mentioned, nausea, or actual vomiting, and not infrequently haemorrhages from various orifices, as a result of the action of the poison on the blood altering its composition, reducing its coagulability, and dissolving the red blood cells.

An easy aid to remember the essential action of the poison is supplied in the word COBRA. CO stands for Cord and BRA for Brain, implying that it is the central nervous system that is in the main affected. Again COBR stands for Coagulation Of Blood Reduced, and the final A gives the mode of death, *viz.*, by Asphyxia.

Cases of cobra toxaemia are very seldom well reported even by the medical profession, a great deal being often left to the imagination.

Rapidity of death in the human subject.—The interval that elapses between a cobra bite and the death thereby occasioned varies considerably. The shortest interval that I have any record of is half an hour. Fayrer reports one case that died in this short interval, the bitten subject being an adult man. *The Pioneer* of the 27th of April 1908, reported a European lady, Mrs. Cockely, succumbing to the bite of a cobra in half an hour. The wound was inflicted on the top of her toe, and the snake was killed there and then by her husband. More often the interval that elapses amounts to hours, from about two to six hours being perhaps usual. A woman mentioned by Fayrer died after 8 hours, and other cases have been reported exceeding 24 hours.

Treatment of cobra bite.—From the voluminous literature on experimental work, with the object of testing various reputed antidotes to cobra or other snake venoms, and of testing the value of mechanical contrivances of checking the absorption of these poisons, one cannot escape the conviction that there is only one known remedy, viz., antivenine. Fayrer's work alone is convincing enough and he spared no pains and gave every possible method a fair trial. Drugs of all sorts, those vaunted by professional snake-men, as well as those from the British Pharmacopiaea were administered by the mouth, by injection into the tissues at the site of the wound, and introduced into the veins, with no benefit. The actual cautery, strong corrosive liquids locally, and the introduction of oxidizing agents such as permanganate of potash, which are known to neutralize and destroy the poisonous properties of cobra venom in a vessel were employed at the seat of the wound without avail. Ligature, excision and amputation were all tried, and proved futile, and the so-called "snake stones" were as useless as everything else.

Fayrer's experiments have been repeated and supplemented by numerous conscientious workers in this field, and abundantly confirmed. The resuscitated "remedy", permanganate of potash, has lately been the subject of an exhaustive investigation at the hands of Colonel Bannerman at Parel and has proved to be completely unsuccessful. The conditions of an accidental bite were imitated as far as possible, and I had the privilege of witnessing some of the experiments. A syringe charged with the lethal dose of poison was fitted on to a Russell's viper's fang. A puncture in the dog's skin was made with a knife point, the fang introduced, and the poison injected. Within a couple of seconds or so, the puncture was cut down upon and permanganate crystals well rubbed in. There was no doubt of the thoroughness of the attempt to bring the salt into relationship with the poison, but it signally failed to avert death. I saw the previous day's dead subject also dissected, and the typical effects of the poison were seen to have been diffused as high up as the thigh, though the envenomed puncture had been made in the foot. These experiments merely confirmed those made by Vincent Richards, who performed no less than 100 operations in the early eighties of last century, and those of Fayrer reported in 1882.

After reading and studying a copious literature on experimental work, there seems only one conclusion to be drawn, and that is, that no method of procedure, whether prophylactic, symptomatic, or so-called antidotal, will avert the fatal issue in cobra bite, where the dose injected is supralethal, except the injection of antivenine. One might, I think, discard the consideration of treatment altogether in cobra bite cases where antivenine has not been injected, and arrive at a faithful estimate of the percentage of fatalities.

Antivenine.—The first steps towards the discovery of this antidote appear to have originated with Sewall in 1886, who proved that an animal could acquire a tolerance for snake poison, till a dose in excess of the ordinarily fatal one carried no ill-effects. Kaufmann in 1889, Kanthack in 1891, Phisalix and Bertrand in 1893, Calmette in 1894, and Fraser in 1895 confirmed Sewall's results, and Fraser succeeded to the extent of conferring on rabbits a toleration to 50 times the usual minimal lethal dose. Calmette, and Phisalix and Bertrand in 1894 and Fraser in 1895 proved that the serum of an immunized animal possessed antidotal properties and Fraser called the product antivenine. In Kasauli where the antivenine is prepared for issue in India, the horse is immunized, and when accomplished to a high degree, the animal is bled and the serum separated. One cc. of the serum as issued is capable of neutralizing 1 mgm. of cobra venom, and the dose recommended for injection into the veins by Lamb and Martin is at least 10 cc.

Antivenine has been experimented with on the lower animals, into which a known quantity of cobra venom in excess of the minimal lethal dose has been injected, and the animals have been saved from an otherwise inevitable death.

The three conditions laid down for successful treatment are—(1) the injection should be made as soon as possible after the bite; (2) it should be made intravenously; and (3) not less than 100 cc. should be injected. At least ten times this amount would be necessary to protect the bitten subject, if injected into the tissues, and there is no comparison to the speed of absorption in this compared to the intravenous method.

The antivenine now issued is reckoned to retain its virtue for a period of two years at least, after which it should be rejected as of dubious efficacy.

Although cobra venom does not directly depress the heart, other influences are very likely to affect that organ in cases of cobra poisoning. Pain, fright, and cold are all powerful depressants to cardiac activity, and may seriously endanger life. It is most essential therefore in treating cobra poisoning to look for any tendency to faintness, and treat this vigorously. Antivenine of unquestionable activity, administered intravenously in adequate doses cannot be expected to save a patient who is suffering from cardiac weakness due to non-toxic causes. A feeble or rapid pulse, with cold body surface, specially noticeable in the extremities, and a subnormal temperature are, though silent, vociferous appeals from an inarticulate subject for vigorous stimulation of the heart. The non-professional attendant can do much in such cases. He can apply friction with powdered ginger or mustard to various parts of the body in turn, whilst the rest of the body is covered up with blankets, and can pursue this course until eight or a dozen hot water bottles can be filled, when they should be wrapped in flannel garments or blankets, and applied all round the patient. He can give hot stimulating drinks, such as coffee, Bovril etc., if the patient can swallow. These should be given in small quantities (half a coffee-cupful or so) every ten minutes. Alcohol should not be given. He can further seek to gain the patient's confidence, allay his fears, and reassure him as to his fate. Possibly the neglect of attention to syncope has been responsible for the disappointing results of antivenine.

■ ■ ■

A Month in the Kazinag Range

By Lieut Col R S P Bates, M.B.O.U. (Indian Army)
Year of publication: 1942

Lieut Col R S P ('Pat') Bates, died in 1961. An ardent lover of birds and a knowledgeable and painstaking field ornithologist, he made significant contributions to Indian ornithology. But perhaps Bate's chief accomplishment and pioneering contribution lies in the field of bird photography in India. Up to the time he published his popular series on *Birds Nesting with a camera in India* in the *Journal* (1924), bird photography here was a neglected area.

Thanks to the unsettled times in which we are now living, a month's leave was all I could obtain. The problem which arose therefore was which would be the most profitable area to work and the quickest to reach in the time available. Finally my wife and I fixed on the Kazinag range, as records of what birds those mountains contain are noticeably few and far between—not that we expected to find anything startling, but it appeared to afford a good opportunity of extending our knowledge of the distribution of Kashmir's birds.

The Kazinag is that range which starting at Baramullah closes the western end of the Vale of Kashmir, providing a stopper as it were to the monsoon between the Pir Panjal and the ranges enclosing the Kishenganga Valley. The north-eastern slopes are drained by numerous short and charming torrents going to swell the Pohru river which itself takes source in the extreme northern limit of the range. These slopes are clothed from almost valley level to between 7,000 and 8,000 feet in magnificent deodar forests which, except for a stretch along the Kishenganga Valley and in the Lolab, occur extensively nowhere else in Kashmir. Amongst the deodars there are of course other trees but this beautiful cedar predominates. At about 7,500 feet where they cease the forest assumes a more mixed character gradually becoming almost exclusively coniferous again as the 10,000 feet level is approached.

The birds of the deodar zone are mainly those of the same levels anywhere around the main Vale and its side valleys, but certain species are scarce–if not quite absent; it is hard to be sure on a month's tour—while others are more common than elsewhere; and one, the Slaty-headed Paroquet (*Psittacula himalayana himalayana*) has his headquarters exclusively in this zone, though parties raid out into the open valleys and spread further afield after the breeding season is ended and the maize has ripened.

We spent the night of June 3 at Baramullah, heavy rain having rendered the fair weather motor road to Handwara quite impassable.

Before leaving the next afternoon in a ramshackle bus whose front springs eventually proved unequal to the strain imposed upon them by that dreadful road, I noted that Slaty-headed Paroquets were not uncommon in the gardens though I only heard and saw single birds. That night we camped where the derelict bus had deposited us, in the vicinity of the bridge over the Pohru river two miles short of Handwara. Shortly before dusk I noticed two huge flights of Jackdaws which appeared to be heading for Sopor.

Throughout the hours of daylight Daws were to be seen carrying food to holes in almost every walnut and chenar tree, so what birds went to make up these enormous evening flights I cannot imagine.

Our objective next day was Sanzipur 11 miles away, but I blush to call it a march for we annexed an ancient tonga, the driver affirming that he could put us three miles upon our way. Thanks to his dexterity in negotiating a much-damaged culvert, he eventually deposited us right at the gate of the Forest Rest House. The excuse for our ultra laziness was to be found in the presence of a Golden Cocker pup for whose legs exercise was still taboo. Needless to say he ran quite wild in the Sanzipur woods, covering many more miles in one hour on his fat little feet than we ever did in a day.

Jackdaw *Corvus monedula*

As we broke camp I noticed Paroquets were numerous, for small flocks appeared quite frequently zooming across the river with harsh squeals to the cover of a bagh near by. The bright yellow terminal third of the tail would be a perfect distinguishing mark were any needed, but the Slaty-headed Paroquet happens to be the only member of the Psittacidae occurring within Kashmir proper.

Although Sanzipur is but 500 feet above main valley level, differences in its birdlife were quite apparent. Most of the Valley birds one would expect to see did occur, but some of these were already rare while a few quite common species were missing altogether. I saw no House-Crows, Starlings, Kingfishers or Paradise Flycatchers. Orioles were uncommon and Tickell's Thrushes very scarce. On all the side rivers running into the main valley from the north I have always found Sandpipers exceedingly numerous from the moment the slack waters are left behind right up to 10,000 feet and even higher. On these Kazinag torrents however, many of which are of fair size with plenty of low bushy cover along their stony margins and divided by many suitable islands, I saw one bird the whole trip, at Kiterdarji on June 27. On our return march to Baramullah as we dropped down to the stream at Panzal I did however hear the unmistakable querulous chittering of these birds being wafted up the hillside on the warm air of the lower valley.

But to go back to Sanzipur. I was struck by the appearance of species normally connected in one's mind with higher altitudes. Meadow Buntings (*Emberiza cioides*) occurred in small numbers in suitable areas, although this is the elevation where around the not so

distant Wular Lake and on the rather arid slopes on the northern rim of the main vale, Stewart's Bunting (*Emberiza stewarti*) holds the field, the Meadow Bunting being quite absent. I had always considered the former as breeding up to about 6,000 feet only and then having its place taken by the Meadow Bunting. In fact I have often stressed that during the breeding months the habitats of these two species do not overlap. Yet at the end of June near Chak-i-Lal Singh, 5 miles from Baramullah where one crosses the last spur, I came upon both species in the same area, on grassy bush-dotted slopes not more than 200 to 300 feet above the valley. The presence of Meadow Buntings is often first given away by their very subdued mouse-like squeaks which they seem to keep up when feeding through long grass and concealing cover, perhaps by way of apprising each other of their whereabouts or of giving warning.

There were a few pairs, very few, of Dark Grey Bushchats and I also on occasion heard the Pale Bush Warbler (*Cettia pallidipes*) well within the forest though there was generally a parrottia dotted clearing of sorts not far off. This shy little bird is not uncommon within the State *rukh* at Achhibal so there was nothing very startling about its presence at this elevation; a point to note however is that in its higher range it seems definitely to avoid forest.

I disturbed a pair of Yellow-billed Magpies in the wood by the Rest House and came upon a Blue-headed Rock-Thrush there seated upon 4 fresh eggs. Cinnamon Sparrows started half a mile further up the valley at Vihom village. But what did surprise me was the scarcity of Hume's Lesser Whitethroats in country which seemed eminently suited to their habits. It is not many miles to those slopes around the Wular Lake where this species is so very common, but except for one pair seen at Sanzipur—I later found their nest in a rose bush—I never saw this bird in the Kazinag. Of course some birds were very numerous—Jackdaws, Mynas, House-Sparrows, and last but not least Slaty-headed Paroquets. It was evident however that we were at an intermediate elevation where species were many but individuals rather few.

Himalayan Wood Owl *Strix aluco*

On June 10 I had my first dealings with a Paroquet at its nest-hole, but that was a red-letter day for other reasons. Three times in the same locality, a deep shady nullah running up into the forest directly behind the F.R.H., we had come upon a Scully's Wood Owl (*Strix aluco biddulphi*), once on a pair of them. I could however find no nest although they seemed rather bold and, when accompanied by the Cocker, we were followed by one of them for about a hundred yards. I decided therefore to try a ruse which seldom pays. I screwed the stand on to the reflex, put in a filmpack, and fixed the strap round my neck. Walking quietly through their usual haunts I drew an absolute blank, but in spite of my awkward burden I fortunately did not give up hope. Crossing the little stream I tried a particularly gloomy patch of forest on the opposite side of the nullah. Quite suddenly I realized I was within twenty yards of an Owl roosting on a lower branch of a deodar and not eight feet from the ground. I approached quietly, not daring to look at the bird lest it take fright. When about 25 feet away I spread the tripod legs and started to focus;

Slatyheaded Parakeets *Psittacula himalayana*

none too easy in so poor a light. However I stopped right down to make sure and exposed a couple of films giving one 5 and the other 15 seconds, during which the bird stared fixedly at the lens but with no undue signs of fear. Moving forward a few paces at a time I exposed a couple of films at each halt until I had used up eight and was hardly 12 feet from the bird. After the last exposure it seemed to have become so used to the procedure that it turned away its head, yawned deliberately, and closed its eyes. Emboldened by this extraordinary behaviour I moved still nearer, but when I looked down into the focusing screen the branch was bare; the owl had taken its departure as silently as a ghost.

At the bottom of the slope the Blue-headed Rock-Thrush was sitting on her eggs. As usual she allowed me to stand right over her, merely putting on that strained look so typical of the Thrushes when disturbed incubating. Unfortunately there were a couple of twigs and some leaves in the way so I could not repeat the Owl performance and photograph her without the hide.

It was still quite early so we climbed half a mile up the wood to a point where I had remarked signs of activity among the Paroquets (*Psittacula himalayana*), so much so that I already had a strong feeling that these birds were perhaps colonial in their breeding habits. Actually I had only marked down one hole which I suspected to be occupied but this quarter of the wood seemed always to be tenanted by Paroquets talking confidentially to one another. This was a circular hole considerably larger than that of a Woodpecker about 18 feet from the ground in the trunk of a deodar.

As we approached I was thrilled to see the head and shoulders of an adult Paroquet protruding from it, so we quickly erected the hide. The slope of the ground was so abrupt that I got the camera comparatively close but even so the lens was tilted at such an angle that the photos I got that day were mere silhouettes against a cloudless sky. The female came back quite soon and stood for some minutes on the perch near the nest and then sidling along the branch levered herself into the hole with the aid of her beak. A woodcutter I sent up the tree said he could make out the edges of two eggs about eight inches to a foot down. On our way back through Sanzipur a fortnight later we could not make out what the hole contained, but the female seemed still to be sitting. This time we built up a platform and erected the tent upon it in the evening; partly so

that I could get to work early before the sun got too high in the sky, and partly because birds seem to be less frightened of a new structure erected in a fading light and become quite accustomed to its presence during the hours of darkness. On the way back we faked a dummy hide 7 feet from the Blue-headed Rock-thrushes which now had three young ones in place of the original 4 eggs, and pushed a large black fir cone into the front to represent the lens.

I returned to the Paroquets at 9.30 a.m. They seemed to be getting used to us, for the sitting bird did not leave the hole until we had been there some minutes and I was almost ready to get into the hide. She was back on the perch very soon after I had been left alone, giving me time for only one exposure before she clambered in. I've called her the female but there was really nothing to go by as to which parent it was.

And now ensued a long wait. My patience started to give out. Nothing happened. The wood settled down to utter stillness but for the tapping of a woodcutter's axe in the distance, and at last I decided to call up the shikari to shift the sitting bird. Fortunately I hesitated, for a few seconds later my boredom turned to feverish activity. With husky screams a band of Paroquets came twisting through the wood; the next moment my startled gaze beheld no less than five paroquets on the perch and clustered round the hole. There were three adults with bright yellow tail-tips and slaty faces, and two youngsters

short-tailed by contrast and a uniform dull green all over. Were they a neighbouring family escorting the head of the house to his front door? The lady within evidently did not care a hoot who they were and took not the slightest notice. In fact, to begin with I was not sure that she had not slipped away unawares but about ten minutes after the roysterers had taken their departure as noisily as they had come, she poked her head out of the hole, had a look round and slowly withdrew to continue her duties.

First, one of the youngsters had a look at the entrance hole, upending himself on a small branch above it and craning his neck to get a better view. But soon one of the adults took possession; perhaps he was the owner. At any rate he seemed the odd man out of the five. I was quickly at work using a so-called silent shutter, which in actual fact made just sufficient noise to attract the audience's attention without frightening them. I was able to take 7 photographs employing short time exposures, as the scratchy noise of the shutter and the slight sounds I made changing plates merely caused them to freeze in grotesque attitudes. Eventually they dashed off for no apparent reason just as suddenly as they had appeared.

To keep all my observations on this bird together I will relate here the remainder of my contacts with them. When we left Sanzipur we first went to Bungus and then back to Nildori. Now Nildori is approximately 8,400 feet. In that locality and at that elevation I never saw a paroquet. It was not until we

White-browed Blue Flycatcher *Ficedula superciliaris*

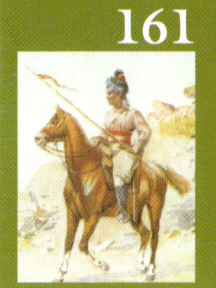

Blueheaded Rock Thrush *Monticola cinclorhynchus* (Male)

dropped down into the deodars again that we met with them. Unfortunately I am not sure of the elevation at which these trees recommenced but it was below the 8,000 feet mark. There certainly seems to be a connection between the distribution of the deodar and the Slaty-headed Paroquet in this part of the Himalayas. After the breeding season they certainly spread further afield. Osmaston records them at Gulmarg in August, and Colonel Ward states that in the autumn bands of them visit the side valleys, Sind, Lidar etc. This wandering is undoubtedly correlated with food supply. In many different villages I questioned cultivators about them and always received the same information. They do little damage to the rice but much to the ripening maize and this of course ripens in the late summer and early autumn. They also attack fruit and in the spring the apple blossom.

I had my final proof of their love of company on July 2 when on our way back to Baramullah. At Panzal, 9 miles there–from and only a few hundred feet above the level of the main valley, the patch skirts the lowest rim of the forest. As we left Panzal we cut

through a little tongue; it amounted to what in the Plains would be called a bagh and consisted mostly of chir pines with about half a dozen very tall birches scattered amongst them. The latter were tenanted by colonies of Slaty-headed Paroquets—unfortunately I omitted to inspect the pines. On the side facing me of one birch I counted five holes, three of which, definitely occupied ones, were hardly two yards apart. Many of the holes were far from being perfect circles, a few of them being somewhat unsymmetrical ovals with the axes at no particular angle. They varied in size considerably; from little bigger than that of a Pied Woodpecker through which the birds squeezed with difficulty to about 4 inches across. The lowest was about 20 feet from the ground while some I estimated to be 60 or more feet up. I saw three or four birds entering and leaving holes while all the time there was an incessant screaming from birds conversing with one another or from small bands weaving swiftly in and out of the trees. These bands contained numbers of immature birds and seemed to be family parties whose nesting for the year was over and done with. Watching them was intensely interesting but alas we were on the move and could not afford to stay.

After that digression I must return to the description of this trip in its chronological order. As there seemed to be nothing at the right stage to photograph, on June 11 we decided to move from Sanzipur to Bungus, a large *marg* lying at approximately 9,700 feet under the main slopes of the Kazinag. As the Naugam F.R.H. was but 2½ miles up the valley, we decided to leave it out and do the march in one. Within less than a mile we crossed a wide almost dry *nullah* bed and walked past the little village of Vihom where I at once noticed that the sparrows flitting about the hedges were cinnamon-headed. It is extraordinary how at times new species appear in numbers with complete suddenness. Entering the shady woods at Naugam we still followed the course of the river for another mile before turning abruptly up the hillside. Birds were few amongst the deodars and little of interest appeared until we suddenly reached the crest of the spur enabling us to look into a broad amphitheatre dotted with widely-spaced pines all of which seemed to have been blasted by lightning or killed by disease; rather a striking sight, especially as soaring majestically heavenwards with but a tiny wisp of cloud trailing from its summit to offset its beauty, there floated over the bowl's further rim the

Blueheaded Rock Thrush *Monticola cinclorhynchus* (Female)

snow-white outline of Nanga Parbat, the unbeaten giant of the Western Himalayas, many miles distant but none the less an amazing sight. To our right I heard the flutelike triple whistle of a Black and Yellow Grosbeak, a bird whose call is as attractive as its brilliant plumage, while the Small Cuckoo was abjuring 'our smoky pepper' from a distant perch.

Following the crest of the ridge in order to reach the first pass into the little valley where lie a couple of Gujar huts and the charming *marg* called Nildori, I suddenly realized we had left the deodars behind and had entered a new bird world. We would then be about midway between 7,000 and 8,000 feet up. Chocolate and white Spotted Nutcrackers

became very common indeed but unfortunately pair after pair seemed to be accompanied by their young. I investigated two nests, both 30 to 40 feet from the ground, which agreed well with the description in Volume I of *Nidification*; but the young had obviously left them days before, so I think their nidification must end normally by the end of May. This bird is far more numerous throughout the Kazinag, from a little beyond 6,000 feet up to at least 10,000 feet, than anywhere else in Kashmir, and I am equally sure that no other bird in that State produces more extraordinary noises. Their calls, if one may use that word, range from wheezy crowlike caws to piggish squeals which they sometimes produce unceasingly for minutes at a time while concealed in the summit of some dense fir.

Tree-Creepers, Tits, Flycatchers, and Grey-headed and MistleThrushes, now became common. I heard once more the sweet rapid song of the Kashmir Wren, the agitated chucks of the Pied Woodpeckers, and began to note many birds of the higher elevations, but I am sorry to say that when at length we attained the last summit and looked across the great stretch of the Bungus *marg*, we were very disappointed. Rainfall had been excessive and the unbroken stretches of green showed deeper patches of colour where the ground had become little more than swamp so that before we reached the tiny, and I regret to say decidedly draughty, forest hut, we were soaked to the knees, so water-logged had it become. As far as the eye could see sheep dotted the landscape, hundreds upon hundreds of them, not only keeping the grass short-clipped but producing a flowerless land where we had hoped for colour and beautiful blooms. Dwarf irises and a few frittilerias were alone to be seen. The surrounding woods were rather thin with acres of fallen trunks on the flanks of the steeper spurs where the crushing weight of slipping snow had battered them down. There and then we decided to keep on the baggage ponies, stay at the hut for one complete day and then retrace our steps to Nildori which looked such a charming little spot and certainly far more hospitable—after all we were on leave and supposed to be enjoying ourselves.

Redbreasted Flycatcher *Ficedula parva*

The following morning I found a White-capped Redstart (*Chaimarrhornis leucocephala*) sitting on a nest of 4 eggs. It was really somewhat conspicuously placed in the upturned end of a fallen trunk close to an almost dry stony nullah running into the woods behind the hut. The bird sat tight allowing us to inspect her at close quarters. I also found a Wren's nest tucked into a crevice in the bark on the underside of a fallen tree and so well concealed from view that it was quite impossible to bring the camera lens to bear upon it. I stuffed up the entrance hole with my handkerchief and then focused on a fallen stick which the bird used as its final pitch on its way to the nest. I did not get good results, however, as the light was too poor to enable me to cope with its very quick agitated movements.

On June 13 we retreated the few miles to Nildori and pitched our tent at the extremity of the flattened spur at the northern end of the narrow little *marg*. It was a charming little spot, well sheltered on both sides and behind by woods carpeted with sky blue Jacob's ladders and with a clear view of three 14,000 feet peaks directly ahead. For the time of year these had an inordinate amount of snow upon them.

Himalayan Tree Creeper
Certhia himalayana

We were glad to have come down from Bungus. I have described the Kazinag range as a stopper. Almost every afternoon clouds from the direction of Uri would attempt to sweep over their crests into the main valley, and alas we struck a period when for the first few days of our life in tents they succeeded in doing so. During the second afternoon we had a terrific downpour accompanied by perfectly deafening thunder. We afterwards learned that it had been the cause of depriving Srinagar of its electricity for a couple of days.

On our way up to Bungus I had noticed a pair of Tree-Creepers (*Certhia h. himalayana*) obviously with young in a narrow slit in the back of a deodar about a mile from camp. The nest was not more than 7 feet from the ground and close to the path; so off we set at the first opportunity burdened with cameras, the hiding tent, and our tiffin. In direct contrast to a very bold pair I once photographed at Sonamarg, both the parents proved shy. It was an hour before one of them plucked up courage to poke more than its head into view and slither rapidly into the hole with a scratchy sound of its thin claws over the intervening bark.

Slither is the correct term and they slithered about the trunk in such rapid jerky little runs that I found it difficult to get off exposures except by using a large stop and a very short exposure. I have certainly taken better photographs but they show off the close barring of the tail feathers of this species very well.

After about two hours work on the Tree-Creepers, we moved back along the path to an old stump with many rotten little cavities in it. One of these held a minute mossy nest of the White-browed Blue Flycatcher (*Ficedula superciliaris*). This is perhaps the most common Flycatcher in the Kazinag and occurs everywhere in the forests from their lowest levels. I also found two nests in quite small orchards at not more than 6,000 feet. The males are easy to identify in spite of their rather unobtrusive habits as their colour pattern of dark blue and purest white is so distinctive. The upper parts appear blue in their entirety except for the conspicuous white stripes over the eyes, widest and all but meeting on the nape. The blue is continued in a broad collar on either side of the breast which however is always interrupted in the centre to a greater or lesser extent. It is not given to conspicuous flights after insects from a fixed perch like the dull-coloured Sooty Flycatcher but its subdued oft-repeated attempts at a song serve to notify its presence whenever one is sufficiently close to hear them. 'Te-che-prrr' it says, dropping its voice considerably on the 'prrr'.

This couple proved easy subjects with the lens but 7 feet from the nest, coming back within five minutes of my being left alone, the female to brood her microscopic babies, the male to bring food both for her and the chicks. He fed his wife at the nest, apparently relieving her of any necessity of forsaking her charges at such a tender age. On his first visit he presented the very aspect I most desired to portray, showing off his interrupted blue breast band to perfection. Unfortunately to let them get used to the lens and hide I let both of them put in a couple of visits before starting photography, and this turned out to be the only time on which the male did not present his back to me.

On the way home I watched a couple of pairs of Mistle Thrushes but could find no nests. The Mistle Thrush is a bird of particularly wide altitudinal range in Kashmir, in the breeding season living in the forest areas from their lower levels at about 6,000 feet upwards and gathering in the late summer and autumn into flocks which find their way to the very limit of the birches at between 12,000 and 13,000 feet. Near the top of the ridge sweet smelling cream columbines were growing.

After this we spent most of our time much nearer home. The wood behind camp was of a very mixed character with clearings here and there dotted with old stumps, bushes and a little juniper. There were also a number of short steep-sided ravines with little trickles and a few muddy patches in their beds. Such country naturally held many species of birds. Tits were scarce, rather to my surprise, but woodpeckers whose nesting was over, abundant. Grey-headed and Mistle Thrushes were common. I found three nests of the former, one very conspicuously wedged in the first fork where the main trunk of a tree divided about 12 feet from the ground, the other two being, as is so often the case, well hidden in the crowns of young firs. Amongst the bushes in the clearings I found nests of Meadow Buntings, two Jerdon's Hedge-Sparrows', a somewhat loose structure of dead grasses. Willow-Warblers of course abounded, particularly Ticehurst's (*Phylloscopus proregulus*), which I learned to trace to its almost invisible little nest in the ends of the fir branches by listening for its sharp mono-syllabic 'Tsip' and then watching it carefully to its lair.

Mistle Thrush *Turdus viscivorus*

The Hedge-Sparrows' we left alone as we had one on an outer branch of a fir not 20 yards from the tent door. I photographed the female on this. It was a beautiful structure, a foundation of quite stout sticks padded with quantities of moss with a thick felted lining of wool, hair, and a few feathers. The whole was intermixed and covered with a layer of a stringy almost white lichen. The female was most amazingly tame and sat on while we raised the sheltering branch immediately above and tied it back out of the lenses', vision, after which I took long-time exposures of her with the silent shutter.

How many people realize that Jerdon's Hedge-Sparrow is no mean songster, its lay being not unlike a Wren's but not so boisterous and penetrating.

Late one afternoon we sauntered along the flank of one of the short ravines. As we neared its head the forest closed in, so, finding it rather gloomy in the fading light we crossed over to return on the other side. It was then that we had one of those outstanding experiences of a lifetime, something to be looked back upon in years to come as an episode never to be forgotten. As we made our way through some bushes on the edge of a more open grass-carpeted slope there arose a strange croaking. First I thought it the production of Nutcrackers but it came from the ground hardly 20 yards away. We called in the Cocker Spaniel and pushed forward. I was rapidly coming to the conclusion that a hawk or owl had got some luckless victim in its talons, when my wife shouted that two large birds had left the ground just in front of her and were flying downhill towards the bed of the nullah. I emerged from the bushes just in time to see a pair of Woodcock (*Scolopax rusticola*) wheel into some cover near the boggy stream below. But where was that easy almost owl-like flight? They flew heavily with tail spread wide, seemingly weighed down behind. And they kept up that strange croaking even after pitching into cover.

We tied up the dog and I started down the slope. Suddenly croaking broke out anew hardly 15 yards below. One of the birds had returned while my attention had been taken up with the dog. I saw a Woodcock on the ground. Its wings were spread wide with the tips drooping into the grass, the tail fanned out, and the body upraised at about 45° with the bill pointing downwards. Croaking and groaning harshly as I rushed towards it, it started into flight, but it was evidently unable to use its legs for the take off so that it dragged down the slope for a yard or two beating the earth with its wingtips before getting clear. Even while flying its body remained at the same awkward angle and it rose but a few feet off the ground coming heavily to earth 30 yards further down.

Woodcock *Scolopax rusticola*

This time I made a still greater effort to catch it in an attempt to elucidate the cause of its strange antics. I could have laid it low with my khud stick, but it just got clear as I stretched out a hand to seize it and succeeded in crossing the nullah where I lost all trace of it.

I think it is now taken as an established fact that Woodcock do carry their young for two purposes, namely to get them to distant feeding grounds and to remove them from danger. I have read various accounts at different times but there are still those who contend that a point still at issue is the actual method of carriage. Unfortunately in this instance the birds throughout were facing away from me, making it quite impossible to see not only the method of transportation but any burden whatsoever, though of course it was quite obvious that they were carrying something. And does this croaking always take place as a matter of course or was it the sudden entrance of our spaniel which caused the commotion?

Nildori provided so many points to explore that even at the end of ten days. I felt that we knew but few of its treasures and still had half a dozen potential photographs in view. But a very large slice had already been taken out of our month so we felt it to be time to move to fresh fields.

We returned to Sanzipur on June 21 stopping there two nights to cope with the Slaty-headed Paroquets. After finishing with those interesting birds, I moved the tent to the Blue-headed Rock-Thrush's (*Monticola cinclorhyncha*) nest in front of which I had previously erected a dummy hide. It now held three well-nourished young ones in place of the original four eggs.

It is just 20 years since I first attempted to photograph this beautiful but elusive Thrush. Anyone who has come upon a nest and has noticed the outraged demeanour of the female thereon, sitting as if nothing short of a forest fire would force her to leave, would imagine that there could be no easier prey for the bird photographer. Unfortunately there are few birds' nests in front of which one can just put down the camera and get busy. There is almost always grass or leaves or some obstacle requiring removal to give the lens a clear view, and once flushed off the nest I have never yet met a more suspicious nature than is possessed by both sexes of this rather common bird.

On this occasion I used an almost silent 'luc' shutter in the extension box on the front of the reflex, and was gratified to hear almost immediately one of the birds in the parrottia scrub close to the right side of the tent. Suddenly it flew straight to the nest and there I beheld the blue and chestnut male standing in strained silence upon its rim while the young ones clamoured for food. I had set the silent shutter for brief time exposures, not venturing to use the clattering focal plane. I pressed in the release letting it go again almost as quickly as I could thereby giving about one quarter of a second. The slight scratch of the shutter leaves caused him to listen intently so that I dared not attempt to change the plate. Full of suspicion he eventually flew off without feeding his babies and that was the very last I saw or heard of either bird. One slight scraping noise had been sufficient to convince both of them of the worst. Still, although I possess only that one negative of the male, I can hardly complain.

My next encounter with this species came a few days later at Kiterdarji, our last resting place. It is only about half a dozen miles from Sanzipur and about the same elevation. To reach it we crossed a couple more of the charming tributaries of the Pohru river, since we made our way directly across the lower spurs of the Kazinag. The country was lovely; forest glades and park-land alternating with fields of ripe corn, acres of linseed nearly as blue as the sky, over which the larks rose and fell singing their loudest,

while at the lower levels the rice was being planted out to the accompaniment of the monotonous yet fascinating chants of the peasants.

This time the Rock Thrush's nest was balanced on a narrow ledge on a decaying stump. It was quite open to view and close to a cattle track through the forest. When first discovered the female continued to sit on her newly-hatched triplets as still as a rock seemingly as bold as any bird could be, yet to get a glimpse of either bird, even from afar, visiting the nest with food was quite impossible. First I left a dummy hide in place for a couple of days and the real one for the best part of another twenty-four hours. When I got inside I waited patiently for an hour and a half but nothing happened. I had taken the greatest care in entering the hide, getting the shikari to stand close to me while I went in through the back. Halfway through the morning I took a short time off in order to try a new dodge. On my return my wife and the shikari both tucked me in and went off carrying an empty coat between them. Now, can these birds count? Or was it sheer coincidence that the male should fly straight to the nest not five minutes later, and almost immediately afterwards the female, seeing the rim of the nest occupied by her husband, landed on a dead stump midway between the lens and the nest so that she was hardly three feet from me.

I was disappointed with my view of the bird at the nest. He squeezed in too close to the trunk with his back to the camera, so as soon as they had both gone I turned the lens on to the stump. Shortly afterwards the female returned, but to my complete disgust she flew straight to the nest and settled down looking directly at me. I don't suppose she could see anything but I had an awful feeling that she was looking straight through the peephole at my left eye. For half an hour I attempted to emulate the proverbial mouse. Fortunately a noisy individual with some cattle and goats came on the scene and off she went. Twenty minutes went by and then suddenly her image seemed to fill the whole focusing screen. There she was standing sideways on the stump eyeing the hide; in fact a glance at the illustration shows a tiny image of it in her pupil. Cautiously I pressed the release and gave an exposure of a good three seconds. She flew to the nest so I started to change the plate. At once she froze and I perforce did likewise, the result being that for perhaps 10 minutes, which seemed more like an hour, we maintained the most uncomfortably strained attitudes; she with her head twisted sideways looking over her shoulder, and I trying to preserve a precarious balance with my hands full of plate-holders and my feet all over the place, as the camp stool had slipped on the steep hillside—their nests so often seem to be on steep hillsides!

Rufous-tailed Flycatcher
Muscicapa ruficauda

A bird which I found extremely common throughout the lower parts of the Kazinag forests was the Rufous-tailed Flycatcher (*Muscicapa ruficauda*). One of the most unobtrusive of the whole family both as regards its coloration and its habits which are really somewhat chat-like, it would often escape notice were it not for the incessant complaining note 'Peup' which it utters incessantly when one is anywhere near its nest. I found we had a typical nest close to the Kiterdarji Forest Rest House, a rather large pad of moss plastered on to the upper side of a fork of a horizontal branch of a fir tree. It was about 15 feet from the ground and 7 feet out from the trunk. From directly below it was quite invisible but from either side it showed up clearly as a substantial compact excrescence considerably larger and far less symmetrical and neat than that of the Sooty Flycatcher. There were three typically spotted young ones in it nearly ready to fly. I had no difficulty whatsoever in getting photographs from the hide placed on a platform laid across two very convenient horizontal branches sprouting from the opposite side of the trunk. I found the parents had varied notes for their arrival at the nest, sometimes uttering a soft 'chur' and sometimes a double 'te-peup', the danger note preceded by a short 'te'.

Red-breasted Flycatcher
Muscicapa parva

Indian Red-breasted Flycatchers (*Ficedula parva*) were common here. In other nullahs I had seen very few, one or two near the Naugam F.R.H. but none at elevations above 7,000 ft. I also saw more Green-backed Tits (*Parus monticolus*) in the Kazinag than I have seen elsewhere in Kashmir. They seem to be very much of a forest bird whereas the Grey-backed Tit is more a bird of the orchards, willow groves, and hedgerows.

Most of these little valleys had a pair or two of Kashmir Rollers (*Coracias garrulus*). One pair were always visiting the Rest House not far from which they caught frogs and large grasshoppers and flew off with their prizes to somewhere beyond the village, about half a mile away. Eventually I found they had three small young ones and an addled egg in a large cavity in a tall willow tree just beyond it. We managed to build up a platform for the hide on the very summit of an adjacent apple tree.

They were bold birds but not half so noisy and acrobatic as their more brightly plumaged cousins of the plains. There were two holes above the nest one of which was very small while a third one, lower down and the largest, directly faced the camera. It soon became evident that the birds had a fixed routine, entering by the larger of the top holes and coming out through the one in front of me. This arrangement did not suit at all as I got a tail end view each time a bird entered and they left without pausing on the rim of the large hole.

After a few attempts I had their normal entry hole stuffed up with a coolie's puggri. The result was quite ludicrous. One of them arrived with a large frog in its beak. For more than ten minutes it alternated between standing forlornly on the puggri and trying vainly to force its shoulders through the small hole near it. Eventually it swallowed the frog and flew away. It seemed quite incapable of appreciating the fact that it could get in through the hole by which it usually came out. At last the other bird appeared. At first it seemed just as much at sea as its mate but all at once light dawned upon it with such celerity that it popped in without giving me time to press the release. Shortly afterwards the male again arrived with a frog., going through the same performance as before except that he flew off to their favourite perch still with the food in his bill. This was doubly unfortunate as they seemed to feed the young ones in strict rotation so that the female promptly refused to come near until she thought he had done his bit. He wasted twenty minutes in flying backwards and forwards before he guiltily scoffed the frog himself whereupon the wife at once condescended to come once more. This time I was the one at fault: she stood so still across the top of the hole that I could not resist taking her portrait with the result that when she suddenly flew down and clutched its bottom rim with her tail spread out across the bark in the very attitude I had visualized, I was in the act of changing the plate. Thereupon feeling that all three of us were being equally futile I went home in disgust to pack up in readiness for the next day's march to Baramullah.

Kashmir Roller
Coracias garrulus

■ ■ ■

Notes on Indian Hawkmoths

By Lieut Col F B Scott, F.E.S. (Indian Army)

Year of publication: 1931

The Hawkmoths belong to the Natural Order *Lepidoptera,* or scale-winged insects. This Order is divided into the Butterflies (*Rhopalocera*), which have the antennae ending in a club, and the moths (*Heterocera*), which have the antennae of various forms other than clubbed at the ends. The moths are divided into Groups, and the Groups into Families, the Hawkmoths or *Sphingidae,* being one of the Families. The scientific name is derived from 'sphinx', the designation used by Réaumur for the English Privet Hawkmoth, on account of the fancied Sphinx-like attitude adopted by the caterpillar when it is alarmed, and the name was later adopted by Linnaeus for the whole Family.

DISTRIBUTION AND FOOD-PLANTS

In India about 180 species are known to occur, out of a total of about 850 species known throughout the world. Some of the Indian species are very common, others so rare that only one or two individuals have so far been obtained. Some of the species are widely spread, others very local in their occurrence. A few of the species which are found in England are found also in India. These are the Convolvulus, Broad-bordered Bee, Oleander, Humming-bird, Spurge, Bedstraw, Striped and Silver-striped Hawkmoths. Two species of Death's-head Hawkmoths are found in India, but they are not the same as the English species. These are the only species which have been given 'common' names. The rest are known only by their scientific names.

Certain parts of India are more rich in species than other parts. Areas with a heavy rainfall and a large variety of trees and plants produce the largest number, and dry areas with poor vegetation the smallest number, though individuals of certain species may occur in vast numbers in both wet and dry areas, becoming serious pests on crops and other plants. Probably many new species remain to be discovered, but of those now known 58 species occur in the west Himalaya (west of Nepal), 128 species in the east Himalaya (east of Nepal, and including Assam), 73 species in the South of India and 40 in Burma. Many more species must occur in Burma, but very little collecting has been done there. The North Kanara district of Southern India is very rich, having over 45 species in an area of 3,600 sq. miles. The distribution of the species overlaps, some of them occurring in more than one of the areas mentioned. The plains area of Northern

Lieut Col Francis Burges Scott was born on 3rd March 1885 and was commissioned into the Indian Army in 1904. He arrived in India in 1906 to join the Regiment of Artillery. Col. Scott's extra-curricular interest was the curious group of insects, the Hawkmoths, which he bred and through intimate observations wrote their life histories and in due course he became an expert and authority on Indian Hawkmoths. He co-authored the Vol on SPHINGIDAE in 1937 in the series Fauna of British India.

Yam Hawkmoth, *Theretra nessus*

SHUBHALAXMI VAYLURE

Acherontia lachesis

India has no special Hawkmoth fauna of its own, but receives contributions from the surrounding areas. The information on this subject is very scanty and it is worth recording the locality where any Hawkmoth is obtained.

The distribution of the moths is dependent to a great extent on that of the food-plants on which the caterpillars feed, though the moths are such fast fliers that they may be found a long way away from the nearest food-plant. Some species feed on a wide range of food-plants, others confine themselves to one or more. The food-plant on which any Hawkmoth caterpillar is found feeding should also be recorded. Plants belonging to the botanical Order *Rubiaceae,* to which *Randia, Gardenia,* Madder, Bedstraw and other shrubs, herbs and creepers belong, is the most popular food-plant, about 30 species feeding on plants of this Order. Vines (Grape vine, Virginia creeper, *Leea*) and Arums (Garden arum, Caladium, Cuckoospit, Snake plant) are the next most popular, with about 16 species each. Leguminous trees and plants (Indian Beech, Shisham, Indian Laburnum, Shiras, Gram, Pulse) come next, followed by Balsams and Spurge. Altogether about 50 Orders of plants are represented in the list of Hawkmoth food-plants known up to the present, ranging from the largest trees to the most insignificant herbs, and including even grasses.

THE HAWKMOTH

I have often heard the questions asked 'What is a Hawkmoth? How can one tell a Hawkmoth from any other kind of moth?'

Amplypterus panopus

In order to do so with certainty, it is necessary to make a minute examination of the veins of the wings and of the organs of the body, but for all practical purposes something more simple will suffice. Hawkmoths can usually be recognised by their characteristic appearance and habits. The fore-wing is long, narrow and sharply pointed ; the hind-wing short and rounded; the eyes large; the chest or thorax heavy; the body or abdomen shaped like a cigar, or like the pointed end of a cigar. This last character is sometimes obscured by lateral tufts of hairs giving the impression of a broad tail, such as occurs in the Humming-bird Hawkmoths and a few other genera; but the other characters are present, and also the clean-cut, high-bred appearance common to all Hawkmoths. Finally, if when strolling in your garden in the evening you notice a moth, poised almost motionless except for its rapidly vibrating wings in front of a flower, suddenly darting away and as suddenly re-appearing, and never settling, you may be sure you are observing a Hawkmoth, since no other kind of moth is known to feed in this manner. If you look more closely you will see that when poised before the flower, the moth unrolls a long tongue or

proboscis, and probes the flower for honey. All Hawkmoths which have been seen feeding, with the single exception of the Death's-head Hawkmoths, have this habit of feeding when on the wing. The Death's-head moths are known to enter bee-hives to steal the honey. The vibration of the wings makes a deep humming note when the moth is flying, and some species produce a similar note when at rest, if they are disturbed.

The Hawkmoth caterpillar can be recognised by the hard, chitinous horn on segment 12, though a few caterpillars of other families have a somewhat similar, but soft fleshy horn.

THE EGG

The eggs of Hawkmoths are round or oval, and are most commonly of a green colour with a translucent appearance, like a tiny, green grape, but they may be almost white or pale yellow, or more rarely brown or orange, in different species. The egg-shell is either smooth and shiny, or dull, and no sculpturing is visible to the naked eye. The eggs of different species vary a good deal in size, the smallest being about 1 mm. in length, and the largest about 3 mm. The average size is about that of a pin's head.

THE CATERPILLAR OR LARVA

The Hawkmoth caterpillar, like other insects, has a head and thirteen other segments. There are different ways of numbering these segments, but we have adopted the method shown in figure 1, counting the head as segment 1, the segment next to the head as segment 2, and so on. The body of the caterpillar is usually round in section, and is more or less cylindrical in some species and in other species increases rapidly in diameter from the head to segment 5, and then becomes cylindrical to segment 12. Segment 2, 3 and 4 each bear a pair of true legs (b). These are hard and shiny, have three joints and a claw for gripping at the tip. They are called 'true legs' as they occur on the same segments as they do in the moth. Segments 5 and 6 are without legs. Segments 7 to 10 each have a pair of pro-legs or false legs (d). These legs are not present at all in the moth. They are fleshy and soft, with a circular pad set all round with curved hooklets, which enable the caterpillar to obtain a firm grip on any surface. Segments 11 to 13 are again without legs, but segment 14 bears the *anal claspers*(e). The anal claspers are similar to the pro-legs but have a still larger gripping surface. So tightly does the caterpillar cling to any rough surface that if it is pulled away, the ends of the pro-legs and claspers are sometimes torn off. Segments 2 and 5 to 12 each have a pair of *spiracles* or breathing holes(c). These, with the air-tubes or *tracheae* which start from them and spread through the tissues of the body, supplying them with oxygen, take the place of the breathing apparatus in mammals, birds and reptiles. The spiracles are oval in shape and have a central slit or opening down the long axis. They are of different colours in different species. Those on segments 2 and 5 to 11 are placed vertically in about the middle of the segments, and that on segment 12 obliquely. Segment 12 bears the horn which is so characteristic of Hawkmoth caterpillars (f). Segment 13 is narrow and rather difficult to make out sometimes, as it is wedged between segments 12 and 14. Just above the anal claspers on segment 14 is the *anal flap*, a fleshy triangular flap which covers the anus. The head is made up of separate chitinous plates fused together into one piece. The front part of the head is called the face (Figure 2). This is made up of a triangular plate called the *clypeus* (a) and the frontal portion of the two lobes (b). The sides of the head are called the *cheeks*. Projecting from the lower part of each cheek are the *antennae* (c). These have three joints, and two bristles, a long and a short one, at the tips of the end joint. The bases fit into sockets in the cheek, and the whole organ is moveable. The function of the antennae is not known with certainty. Between the antennae are the powerful jaws or *mandibles* (d) with their

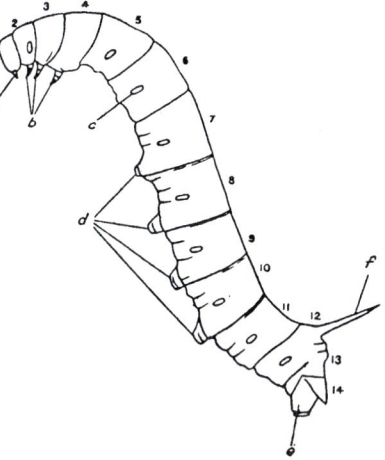

The Caterpillar or Larva

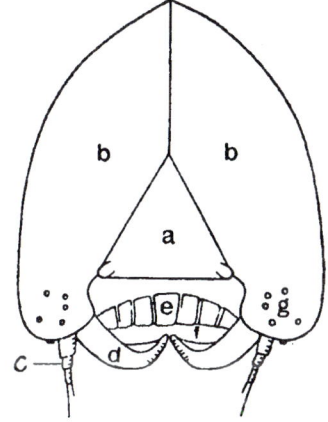

Head of a Hawkmoth Caterpillar

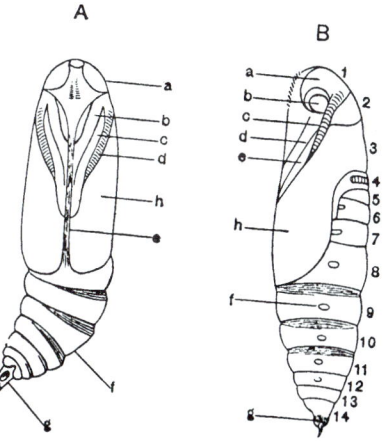

Pupa or Chrysalis of a Hawkmoth

bases also set in sockets in the cheeks. They are wedge-shaped, curved near the tips, with bevelled edges working against each other sideways, and are used to cut pieces from the leaf when feeding. Behind the mandibles are mouth-parts called the *labrum* (e) and the *ligula* (f) which come into play when the caterpillar is feeding. Above the base of each antenna is a group of five eyes directed forwards, and a sixth eye is near the base of the antenna but directed downwards (g). These are very small, hardly visible to the naked eye, but under a magnifying glass appear as circular, convex, black dots. It is doubtful if the caterpillar can see more than a few inches with these eyes. Behind and below the mandibles is a small cone with a perforated tip, which is the *spinneret* from which a thread of silk can be spun at will.

Cephonodes picus

PUPA OR CHRYSALIS (figure 3, A and B)

The Hawkmoth pupa has a shell or casing of hard chitinous material, inside which the moth forms. The pupa has the same number of segments as the caterpillar. The head (segment 1), the thorax (segments 2 to 4) and segments 5 to 8 of the abdomen are all fused together so that they are immoveable, but the remaining segments of the abdomen are jointed and moveable. The head is usually round and blunt, the abdomen pointed as in the moth, the body being thickest in the middle. The case is so moulded that the position of the head, eye, tongue, fore and middle legs, the antennae, folded-up wings and the body of the future moth can be seen. The tongue runs down the middle of the ventral surface, and may or may not reach the end of the wing-cases. On either side of it are the lower part of the fore legs, then the middle legs and then the antennae. The hind legs are concealed under the edge of the wing-cases, which start near the antennae and reach about half way down the ventral surface of the pupa. At the end of segment 14 is the *cremaster* (g). This is an organ of hard chitinous substance, which is either triangular or spike-shaped, and usually branches into two points. These points may again divide into two, and there may be one or more pairs of small hooks. The shape of the cremaster and the arrangement of the hooklets provide a valuable means of identifying different species in the pupal stage. The cremaster does not appear to perform any function except when provided with hooklets. When these are present they are used to fix the tip of the abdomen to a pad of silk woven by the caterpillar at the end of the cocoon. Spiracles are present on segments 2 and 5 to 12 as in the caterpillar, though that on segment 5 is concealed by the edge of the wing-case.

Clanis titan

In some species of Hawkmoth the tongue is very long, and it cannot be accommodated in a pupal case. It is then housed (to use a mechanical term) in a special hollow casing which projects in front of the head of the pupa. The Convolvus Hawkmoth (and a few other species) has such an excessively long tongue that it cannot be accommodated even in this manner, and it is then housed in a free tongue-case. The tongue starts from the front of the head, runs along the case to its bulbous end where it turns back on itself and re-joins the head casing, then runs between the wing-cases to the end of the latter. This free tongue-case looks like the handle of a jug.

In colour the pupa is chestnut or dark-brown in the case of those species which pupate underground, and of various colours, with dark or pale dots, stripes or patches in the case of those which pupate on the surface. The surface of the pupal case may be smooth and shiny, or dull, and is sometimes shagreened or covered with small tubercles, and sculpturing is sometimes present on segment 4 or near the spiracles and head.

LIFE HISTORY AND HABITS

The Hawkmoths, with very few exceptions, lay their eggs singly, usually on the undersides of the leaves of the food-plant or plants on which their caterpillars feed. Each egg is stuck firmly to the leaf or twig with some sort of gum secreted by the moth. The operation of egg laying has not been observed in natural conditions in the vast majority of Hawkmoths, as it is usually carried out after dark, but a few day-flying species have been seen laying their eggs, and they have done it on the wing, without settling. While poised delicately over a leaf or a young shoot, the tip of the abdomen is turned up or down and an egg quickly deposited. The moth then darts away to repeat the operation elsewhere. When a large number of females of any species are engaged in egg laying, several eggs may be found on a single leaf or plant but they are almost certainly laid singly at different times by the same or different females. Most butterflies, and a few moths also lay their eggs singly, but the eggs of butterflies can usually be distinguished from those of Hawkmoths by being of various shapes, and by the shells being sculptured into patterns visible to the naked eye. The eggs of some of the swallow-tail butterflies are very similar to Hawkmoth eggs, and one might be taken in by the resemblance until the young caterpillars hatch out. All doubt is then dispelled by the presence or absence of the horn which is the distinguishing mark of the Hawkmoth.

Daphnis nerii

The egg usually becomes paler in colour a few days after being laid, owing to the transparent shell allowing the colour of the caterpillar which is forming inside to be seen. If the egg is examined under a fairly strong magnifying glass just before the young caterpillar is due to emerge, the head and some parts of the body may be made out.

The young caterpillar or *egg-caterpillar* comes out in 5 to 10 days after the egg is laid, the larger species usually taking a longer time to hatch than the smaller species. The egg-caterpillar eats a hole in the side of the egg-shell, and makes its way out in a minute or so. Most commonly it is of a pale yellow colour, including the horn, but the horn soon becomes black. The body is covered with hairs which are visible to the naked eye in some species. The caterpillar eats the egg-shell for its first meal and after resting for a time along the midrib or a vein of the leaf, starts to eat the leaf itself. It often eats small holes in the middle of the leaf at first, and tackles it from the edge when it grows a bit bigger. After feeding for a few days the body becomes too big, not for its boots, but for its head and legs, which are unable to stretch like the skin of the body does. The caterpillar then settles down to change its skin and acquire a larger head. It lies motionless along a midrib or vein for some hours, and then the new larger head may be seen forming under the skin behind the old head. The old head is pushed forward till the skin breaks round the neck. Then by an undulating movement of the body the old skin is worked back, until, with a wriggle of the claspers it is cast off, and the old head is also got rid of. Some species eat the cast-off skin. After resting for a time the caterpillar starts feeding again, and when the body becomes too large for the head it changes its skin again. In most species there are four such changes of skin before the caterpillar reaches the final or mature stage (each stage being called an *instar*), and there is some change of colouring or form or both at each moult. The ocelli or other markings gradually develop, and the shape of the body and of the horn may differ in each successive moult. The caterpillar feeds more and more voraciously as it nears maturity, then suddenly stops feeding and remains motionless for about 24 hours. During this period of rest it often becomes of a darker colour in preparation for its descent to the earth, green caterpillars assuming a pink or brown suffusion along the back, and dark-coloured caterpillars becoming still darker.

Deilephila elpenor

Hippotion celerio

Macroglossum corythus

Marumba dyras

Polyptychus dentatus

Suddenly leaving the food-plant the caterpillar begins to look for a suitable place to pupate. Those species which pupate underground get very agitated and hurry along the ground with a quick undulating motion, and if touched, lash their bodies wildly from side to side. Their pro-legs and claspers gradually lose their power of gripping, and the caterpillar falls over lumps of earth and other obstructions in its anxiety to get safely underground. When it finds a soft place, it immediately starts digging with its head, and very soon disappears under the surface. It may dig down to a depth of 6 or 8 inches, and there makes a large oval cell in which it turns to a pupa.

The species which pupate on the surface (these are the larger number) do not have to travel so far to find a suitable spot, and are more leisurely in their movements. They crawl under dead leaves and vegetation, and make a rough cocoon by joining leaves, earth and rubbish together with a few strands of silk from the spinneret.

The change to a pupa takes place from two to ten days after going underground or starting the cocoon, but in one or two species, may not take place for several months. The change to the pupa is not carried out in the same way as the moults in the caterpillar. When ready to pupate, the head of the caterpillar splits down the front, and the head of the pupa is pushed through the slit. The skin of the caterpillar, with the head attached to it in two halves, is then worked back over the body of the pupa. The pupa is at first soft and shapeless, and the sheaths or cases which will later on contain the antennae, legs, wings and tongue are separate from the body, but they soon fall into their final positions and become firmly fused to the body. The pupal case hardens and assumes its final form and colouring. After lying nearly motionless for a period of from a fortnight to several months, according to the species and the time of year and other factors, the pupal shell splits open along the dorsal line of the thorax, the head and tongue case breaking away together, and the moth emerges and dries its wings, and darts away to feed and find a mate. After mating, the male dies, the female lays her eggs to start a fresh brood and then she also dies.

This is a short outline of the life history of the Hawkmoth. There are many variations, some common to whole subfamilies or genera, others peculiar to certain species, but it is not possible to give more than a general account in these notes.

COLOUR AND MARKINGS

When first hatched, the Hawkmoth caterpillar is usually some shade of pale yellow or yellow-green, and is without markings. In a few cases only the colour is brown or black. After feeding for a time the green colour of the food sometimes shows through the body, giving it a green tinge. In the second and third instars, that is, after the first and second changes of skin, the colour is usually green, and the oblique stripes and other markings begin to appear. Where the mature caterpillar has the eye-like markings called *ocelli*, these first show as round spots of a uniform colour, and develop with each change of skin till they reach their final form. In the greater number of species the colour remains green in the fourth and fifth instars (the fifth usually being the final instar before the caterpillar pupates), but there is in some cases a startling change in the fifth instar, the ground colour of the head and body changing from green to brown, black or purple. The oblique stripes and other markings may remain unchanged or may be greatly modified. Even in the case of those species in which the colour is normally green till maturity, individual caterpillars may assume this dark form of colouring in the final or in earlier instars, and in a few species there are three or more differently coloured forms. The various forms are so unlike each other in colour and sometimes in markings as well, that one would not believe them to be the same species, but the moths bred from the different forms are

identical. There are other cases where the change to a dark form is not complete, certain individuals developing dark patches which do not cover the whole body. In the few cases where the egg-caterpillar is black, the colouring may remain black (or dark) throughout, or there may be both dark and green forms. The different cases may be summarised as follows, in the order of their occurrence in nature:

(a) The caterpillar is always green in the earlier instars, later has both a green and a dark form, or three or more different forms.

(b) The caterpillar is always green from birth to maturity.

(c) The caterpillar is always green in the earlier instars, later has a green form with or without dark patches.

(d) The caterpillar is always green until the last instar, always dark in the last instar.

(e) The caterpillar is dark in the earlier instars and later has only a dark form, or both dark and green forms.

The occurrence of two or more differently coloured forms in the caterpillar, with no corresponding change in the moth, is very curious. It cannot be accounted for by any difference in food, since the different forms are found feeding on the same plants. The green form is usually the most common in nature, or at least the form most commonly found, but when specimens are bred from an early stage in a dark tin or box, a far larger proportion of them assume the dark form. This seems to show that absence of light is a factor in influencing the coloration. On the other hand, dark coloured specimens are found in nature in the same situations as the green forms, both forms being exposed to the same amount of light. Also, where there is only a dark form at maturity, the dark, mature caterpillars are often found during day-light in the same situations as green forms of other species. I had a curious experience with caterpillars of the Convolvulus Hawkmoth at Sheikh Othman, near Aden. There were large numbers of them in the earlier green stages on a certain creeper, but no mature caterpillars could be found. Someone suggested keeping the small caterpillars in more or less natural conditions in a large box. On doing so it was found that all specimens turned to a dark form at maturity, and that during the day they left the food-plant and hid among dead leaves and even buried themselves in the earth to avoid the light. This experience does not lead us to any conclusion, since it may be argued either way–that the caterpillars hid themselves because they had assumed the dark form, or that they assumed the dark form because they had developed the habit of hiding during the day. Further evidence on this question is required to enable the problem to be solved.

In addition to the general colouring, Hawkmoth caterpillars have various markings, the most common style of markings being longitudinal stripes, oblique stripes and ocelli. Longitudinal stripes may be present in combination with either oblique stripes or ocelli, or all three types of markings may appear together. The longitudinal stripes may be present along the back (*dorsal*), high up on the side (*dorso-lateral*), through the spiracles (spiracular) or below the spiracles (*sub-spiracular*). The oblique stripes are usually seven in number, on segments 5 to 11: that on 11 extending upwards and backwards over segment 12 to the base of the horn. The ocelli occur in one pair on segment 5, or two pairs on segments 5 and 6 or in seven pairs on segments 5 to 11. In one species there is an extra pair on segment 4. The ocelli are round or oval, and usually have a dark centre surrounded by a paler colour and then a dark ring. In some cases they are convex in section and shiny in appearance, and then the resemblance to a real eye is increased. The ocelli usually lie on the dorso-lateral line, but in a few cases the spiracle on segment 5 is ringed with colour, so that it resembles an ocellus.

Polyptychus trilineatus

Psilogramma menephron

At each change of skin, the shape of the head, body and horn may change, as well as the colouring. In the egg-caterpillar the head is always round, and it may remain round to maturity, or it may becomes triangular or pointed in the second instar. In a few cases the head of the egg-caterpillar is round, it then becomes triangular or pointed, and at maturity again becomes round. The body is nearly cylindrical at birth, and remains so to maturity in some species, while in other species the fourth and fifth segments become tumid or swollen. The horn of the egg-caterpillar is always straight, slightly tapering, and *bifid*, or with two points, and each point bears a hair or *seta*. The double point is usually lost in the later instars, but in some cases persists to maturity. The horn may remain straight, or it may become curved downwards, more rarely upwards, and in a few cases it is twice curved, first down and then up, as in the Death's-head Hawkmoths. The shape, thickness and relative length varies greatly in different species, and in some of the genus *Clanis* may be so small as to be overlooked.

In the egg-caterpillar the horn can be moved at will in a vertical plane and this limited power of movement is retained in a few species, where the horn is very thin up to maturity, but in most species all power of movement is lost in the later instars, and the function of the horn, if any, is unknown. In some Hawkmoth caterpillars from South America the horn is very long and whip-like and can be moved freely over the back of the caterpillar like the filaments of the Pussmoth caterpillar. In these species it may serve to drive away parasites, but it is not long or mobile enough to be of any use for this purpose in any Indian species.

The surface of the head, body and horn is usually dull, and either smooth or tuberculate. The tubercles may cover the whole surface, or may be present on certain parts only. Sometimes only the horn is tuberculate, or the tubercles may run along the back or along the line of the oblique stripes. In one species they have developed into long fleshy spines, and in others into wart-like prominences. Hairs are always present, but except for a few on the head, legs, pro-legs and anal flap, are too small to be seen without a lens.

MEANS OF DEFENCE AND ENEMIES

The Hawkmoth caterpillar is a most defenceless creature. Having no long hairs or poisonous spines such as many other kinds of caterpillars have, it falls an easy victim to predators. Its only hope of escaping from its enemies is to avoid detection, and for this purpose its habits and colouring are admirably adapted. It habitually lives on the underside of the leaves of the food-plant, and grasps the midrib or a vein with its pro-legs or claspers, its head directed towards the point of the leaf. When resting it is entirely hidden from above by the leaf, and when feeding only the head and one or more pairs of legs are visible. When young, the pale colouring matches that of the midrib, and later, when the green colour and the oblique side stripes have developed, the latter lie parallel to the side veins, and the whole creature appears to melt into the leaf. Some species have irregular spots which look like dead patches in the leaf. When once discovered and attacked, the caterpillar can only defend itself by raising the front part of the body and hitting sideways with the head. The Death's-head caterpillars increase the effect by making a clicking noise with their jaws, and those of *Langia zenzeroides* and other species make a loud squeak at each stroke by expelling air forcibly through the spiracles. Those caterpillars which have ocelli draw in the head and anterior segments into segments 4 and 5, at the same time puffing out these two segments as to expand the ocelli to their full extent. This gives them a somewhat snake-like appearance, and the effect is enhanced by their raising the fore-part of the body and waving it from side to side. The Spurge Hawkmoth caterpillars have a more effective means of defence. They live gregariously on the Spurge, a dozen or more perhaps on a plant. When disturbed, they throw back the head and fore-part of the body and eject drops of clear green fluid. The effect of their simultaneous action is most startling, and probably serves to drive away small enemies. Their striking appearance, lack of concealment and the poisonous nature of their food-plant point to their being unpalatable to birds.

The enemies of the Hawkmoth caterpillar are many and varied. The wolf-spiders jump on them when they are small, and suck their juices. Ants of most kinds consider them fair game and attack them regardless of size. It is reasonable to assume that birds and lizards eat them when they can find them, and that hunting wasps carry them off to their burrows or cells, though I have never actually seen them doing so. Judging from the gusto with which some Slender Loris I once kept devoured the huge caterpillars of *Clanis phalanis,* such insectivorous mammals must take their toll. They would seize the caterpillar in both hands, and having first crunched up the head would work steadily down the still squirming body, the green juices dribbling down them the while—a horrid proceeding. But the really insidious foes, which probably cause more destruction than all the rest put together, are the parasitic wasps and flies. One minute species of wasp lays its eggs in or on the eggs of the Hawkmoths, probably when the latter are freshly laid and still soft. Small grubs hatch out and eat the substance of the egg. The eggs which have been attacked turn first grey, and then mottled with black and white, the white patches being the pupae of the wasp. If kept under observation a tiny wasp will be seen to come out of a small round hole which it has bitten through the shell of the egg, and the rest of the brood will scramble out after the first at short intervals.

The caterpillars themselves are attacked by several kinds of parasitic wasps and flies, from the time they are about half-grown to maturity. The method of attack is the same as in the case of the egg, but the parasites are much larger. The grubs feed on the tissues of the living caterpillar, which carries on as usual till one day it is seen to be covered with small cocoons like ants' 'eggs', which stick out all over its body like almonds in a pudding. The maggots, being full fed, have made their way out through the skin of the caterpillar and formed their cocoons there. If brushed off, each cocoon leaves a black spot on the skin, and the caterpillar presently dies. Other species of maggots carry on their fell work unseen till the caterpillar is found, still holding on by its claspers and one or more pairs of pro-legs, but the upper part of its body hanging down limp and empty except for the squirming maggots. These when full fed form hard cocoons inside the empty skin, from which the perfect insects emerge in due course. Sometimes the caterpillar succeeds in pupating. The maggots continue their horrid feast, and form their cocoons inside the pupal shell. Then, instead of the beautiful moth we had expected, half a dozen nasty-looking flies or a wasp or two come out. Some species seem to be far more often attacked by parasites than others, and to suffer more in some years than in others.

It is not known what enemies the pupae have in nature, since they are so seldom found, but when kept in artificial conditions they are attacked by a small black fly.

The moths are so swift on the wing, and hide so skilfully when resting, that they probably have few enemies, but bats certainly catch them. A friend once told me that he used to find the wings of Hawkmoths in his verandah every morning, under a hook where a bat used to hang while it devoured their bodies. All observations on the subject of enemies and means of defence would be of interest.

Death's-Head Hawkmoth
Acherontia lachesis

On the Banks of the Narmada

By Lieut Col R W Burtons, I.A. (Indian Army)

Year of publication: 1941

Born in 1868, **Richard Burton**, the sixth son of Gen. E.F. Burton of the Madras Staff Corps, was commissioned from Sandhurst in 1889. Posted to the Indian Army in 1890, he was permanently crippled by a riding accident in 1903. He was, thereafter, assigned to the Cantonment Magistrates Department. A fearless sportsman and a keen fisherman, he wrote over 200 articles on various aspects of Natural History and was the first Naturalist to campaign for the preservation of Indian Wildlife. Col Burton passed away at his residence at Surrey, England in January 1963 at the age of 95.

'Narmada Mai', or Mother Narmada as it is also reverently named, is considered by many Hindus to be the most sacred of all the rivers of India. It rises to the east of the Central Provinces, at Amarkantak on the borders of the State of Rewah, and enters the Arabian Sea near the town of Broach after a course of some seven hundred miles.

In earlier years it formed, with the forests and hills along its course, one of the main barriers which then shut off the peoples of Northern India from those of the Deccan. At the close of the triumphant career of Samudragupta, the second king of the Gupta dynasty, the Narmada was his southern frontier. He did not attempt to retain conquests made south of the river, and returned about the year A.D. 330 past the fortress of Asrigarh which is nowadays seen by railway passengers between the stations of Burhanpur and Khandwa.

The Narmada (Sanskrit, Nar-Mada, 'causing delight') is rightly named for it is a beautiful river through most of its long course, and to camp on its banks during the cold season is truly a delight. In the hotter months of the year the pleasure may be somewhat abated; but tiger and panther are more readily come by, and at certain places the fishing is good.

If one's stay is leisurely suitable swims can be baited with parched gram by means of which excellent sport is obtained. This gram fishing is a method peculiar to the Central Provinces, and an art in itself. Suitable equipment is a fly rod for fish up to 10 lbs. or so. A gut cast as fine as you dare use, and size 8 'model perfect' hooks. 'Fine and far off' is the maxim to bear in mind, and the threaded pellet must swim naturally with the other grains thrown in to accompany it down the stream; for any 'drag' will be fatal to success. And unless there is a ripple on the water it is likely that *all* the pellets will be sucked down *except* that to which one is expectantly attached! It is expert angling, can be most disappointing at times, and is not so simple as the chapter on the subject in the *Rod in India* by the immortal Thomas would lead one to suppose. Anglers wishing to have the fullest possible information and guidance as to gram fishing should refer to the excellent article on the subject entitled 'Fishing in the Rivers of the C.P.' by Maj. W.B. Trevenen which was published in the *Journal of the Bombay Natural History Society*, Vol. xxxiv, No. 3, p. 600 *et seq.*

Where there are boisterous falls in the river, as at Dhariaghat, there is excellent sport, using natural bait, with mahseer and a number of other predaceous fish both for a period when the water is subsiding after the rains and in the hot weather—March being the best month if the intense heat is not objected to. To hit of the former some considerable margin of time is necessary, or there may be disappointment due to late floods. The waiting days can be passed in pursuit of tiger and panther, in natural history observations, or just walks through an always interesting forest country.

One day three large stones had to be moved to one side of a forest track to clear the way for the cart. Under one stone was a snake, beneath the next a centipede quite eight inches long, and the third harboured a scorpion! Trails of large pythons are occasionally seen in sandy places; huge spider's webs are stretched from bush to bush; the night's tracks of all the jungle animals and birds are along the paths: so jungle strolls can be full of surprises to an inquisitive and observant eye.

There is much life on the river and along its sand banks and islands. We see a crocodile on a sand-spit, and perched on the near-by jutting branch of a submerged snag is a Darter, 'snake-like', he appears when his lean head and neck are protruded above the surface of the water after a dive. The specimen we see has his wings spread out to dry and looks like a bird on a lectern—or the German Eagle! At a respectful distance from the seemingly sleeping saurian are two brahminy ducks—Ruddy Sheldrake, to give them their other name. Wary birds they are, and without good reason, as they are not sought after by European sportsmen and are protected by Hindus, who do not like them being shot. Graceful River Terns are seen sweeping over the water with light and airy flight, and kingfishers of three varieties are noticed: the black and white species ('The pied fish-tiger hung above the pool') being less common on this river than the larger and smaller coloured ones—the Common Indian Kingfisher, with appearance and habits of the one so familiar to us in the British Isles; and the Indian White-breasted Kingfisher, with large, conspicuous red beak.

Darter
Anhinga melanogaster

Cormorants are common, and that curious bird, the Indian Stone-Curlew, or Goggle-eyed Plover, is often flushed as we float silently in the dug-out past the islets of the river. Among the bright green foliage of the dwarf jamun bushes many small birds, bulbuls, warblers, sparrows and the like are seen and heard; while the ubiquitos King Crows and their larger and more ornate relatives the Racket-tailed Drongos, are seen hawking the air from selected vantage points. Screeching paroquets weave emerald skeins among the tree-tops; egrets show dazzling white amid the dark foliage of the trees; there is the occasional splash of fish; and a wide ripple is caused by a crocodile having slipped silently into the stream from a sloping bank.

Brahminy Duck (Ruddy Shelduck)
Tadorna ferruginea

Troops of macaques and langurs, an occasional otter; all these and many other sights are the ordinary daily happenings; but all that is to be seen and noticed would fill a volume. Indeed this river is a delight not only on account of the many forms of life to be observed, but because of the changing forest scenery; the gracefully waving bamboos, grasses and tamarisks; the lovely lights and shadows on the water; and the beauty of the

The Temple of Omkar on the Island of Mandhata

crimson sunsets followed by the softening light of the risen moon which makes the wide river bed with the dark brooding forests on either bank a fairy land of magic and mystery.

The rainy season in the north-western part of the Central Provinces is frequently prolonged into October, so when setting out late in September we knew there might be some unpleasant weather and cart tracks in a bad state. However, after two marches to cover the twenty-four miles from the small wayside station we pitch camp within sight and hearing of the falls under the scanty shade of small teak trees from which the leaf has mostly fallen.

Above the basaltic barrier the river is now some four hundred yards in width, narrows to a hundred and fifty, and tumbles in separate cascades of varying volume over the rocks through which it has cut its way during countless centuries. It is a fine sight.

Below the falls the perpendicular rocks confine the river in a gorge eighty to a hundred yards wide, to open out and again contract, and so make its way for twenty miles to Mandhata, where are famous temples and a cliff from the summit of which human sacrifices took place up to the year 1824, when the country came under British dominion at the close of the Third Mahratta War. It is plain to the eye that the falls we now see were, ages ago, some eight hundred yards further down; and aeons before that again, several miles below.

By the side of the path approaching the sacred river is a small temple dedicated to Mahadeo, from within which issues the almost unceasing chant of the solitary priest. All though the day and night, with but short intervals for rest and food, he drones away; and so passes the monotonous days of his earthly existence in contemplation and hope of the Nirvana he aims eventually to attain. He is one of the really earnest devotees and has been here three years, alone in the tiger-haunted forest, with no human dwelling within miles of him on this side of the river.

In earlier days similar devotees were carried off from the jungle temples at Amarkantak by man-eating tigers: and that may be so occasionally even at the present time.

Early on the 4th October we are on the rocky bank below the nearest fall. The water is beginning to clear; so a good sized *mural* is seen cruising about. With us is a *dhimar*, a man of the fisherman caste summoned from the Dhar State on the further bank. Having lived much of his life with the roar of the falls in his ears he has acquired a habit of conversing by signs, so interprets his meaning by taking off his loin cloth which he fashions by means of pieces of driftwood into a sufficiently serviceable net. Squatting at the edge of the rough water he has in a few moment bait for the undoing of *Ophiocephalus striatus*—and whitebait for our dinner! He then takes the small rod set aside for his use, attaches a treble hook to the line, and, stalking the place quietly drops his bunch of wriggling fry into the rushing stream ten feet below. In less than no time the reel is singing a merry tune. 4 lbs. and 24 inches long: one of the best eating freshwater fish of India. But all *mural* are not so simple as that.

On the other side of the river, where the rapids from the main fall sweep in rushing eddies along the rugged rocks, are two men

The Narmada crossing: the Forty Ferry

with sixteen-foot bamboos and twenty feet of cord. Using dead bait they search the likely places with perpendicular lines and we see three large *mural* unceremoniously lifted out of the water. 'Sometimes,' says our henchman, 'they catch as much as three maunds of fish in a morning.' It is mahseer we want and not these coarser fish. The water is too coloured for spinning and we try live bait without result, for the river is rising fast and there must have been heavy rain higher up.

In the evening we go to see the reported tracks of a tiger. A fine fellow he is by his footprints, of which we see both old and new, so know that the sandy ravine is where he passes regularly on his rounds. A tree for a machan chair is selected. There being no suitable root or stump to which the young buffalo can be picketed, a trench eight feet in length and a foot wide is dug, with a further length tunnelled at each end, so that a ten-foot log can be sunk two feet below ground level. Round the log is passed a flexible wire rope with a loop at the surface to take a similar rope for attachment to the poor boda's foreleg. The loop round the animal's leg is carefully padded. The sand being filled in and well rammed down over stones, adding to the weight, all is secure. No tiger can move such a fixture.

A tiger killing at this prepared dinner table should be as good as bagged. There must be no wounding; for, the jungle, apart from any other consideration, is so thick at this season that the following up of a wounded beast would almost amount to suicide: yet it would have to be undertaken for such are the ethics of the game.

Next day a thunderstorm makes everything damp and uncomfortable, but on the following morning a cloudless sky evidences more settled weather. The water is now less coloured and a 3-lb. mahseer is secured on a spinning dead bait, and in the afternoon another of the same size on the further side of the river. Next morning two mahseer of 5 and 6 lbs. are taken, also on spinning tackle, and then the uncertain weather brings the river down in heavy spate. The rain to cause this six-foot rise must have been a long way off and very heavy: probably it was around the head waters on the Rewah border. Here is a cloudless sky, hot sun, and chilly nights. At 6 a.m. it is 66° and 96° at 2 p.m.

For twelve days the river continues in flood. Neither we nor the locals take many fish, and none of them are mahseer. For days our desultory efforts produce only *tengra* of one to three pounds weight. These scaleless fish afford no sport, but are good to eat. Several *perrun*, also of the family Siluridae, are caught on live bait, the largest 8 lbs.— ugly fish having enormous mouths full of rows of sharp teeth. All fish of this 'cat-fish' family are voracious feeders, and some of them grow to a great size—six feet in length, and 200 lbs. and more in weight.

Omkarji: The Sacrificial Precipice

In 1891 I hooked one quite six feeet long from the stern of an Irrawady paddle-wheel steamer lying at anchor for the night below Katha. A lascar foolishly touched the taut line with the boat-hook so the monster escaped. A *wallago-attu—perrun*—of 23 lbs. seized a *rohu* (carp) of seven pounds while the smaller fish was being played in the lake at Mahoba; after he let go a 2-lb. *rohu* used as live bait brought about his speedy capture.

Sometimes we sit and watch the leaping fish. Gallant efforts they make to ascent the foaming falls and cataracts; some must succeed, but most appear to fail. Our greatest admiration is gained by the tiny heroes which leap from the foaming torrent against the slippery sides of the rocky walls. There they cling, panting, with down-pressed pectoral and

A Narmada Fort of Pindari Days

ventral fins; wriggle up a little, take a rest, wriggle again, and so, by difficult inches, climb five feet of rock—mostly perpendicular and sometimes overhanging—until they meet the water trickling over at an edge. Then a final effort is made and the gallant mite rushes into the flood to be instantly borne away down the fall he has been instinctively endeavouring to surmount.

It is only very wee minnows which attempt the ascent in this manner. The others are ceaselessly leaping, leaping, mostly without avail: and the loin cloths are busy all the time.

On the 13th October here was a great gathering of people from far and near, the occasion being the annual Dasara festival—the Durga-Puja of the Bengal Presidency. On this day there will have been similar gatherings at all the ghats and temples throughout the course of this sacred river.

It is a gay scene we look upon. The rugged, rocky shores of the river are crowded on both banks with gaily apparelled men, women and children, for this is a universal Hindu outing similar to a British bank holiday. Bullock carts, palanquins, ponies, are scattered among the throng, and the smoke from many fires rises in the air. There is a great washing of clothes, performing of various ceremonies, and offering of coconuts. Below the falls are several large boats which have brought people from Mandhata, and the boatmen are reaping a rich harvest of the nuts from the stream, vying with each other in racing down the flood to secure them. By evening all these thousands of people have left to return to their homes.

Next day there is news of a large cow buffalo having been killed by a tigress five miles upstream. We hasten to the place with a young companion who is in camp with us for a week or so. Mr. Verdant Green—for so we may name him concerning all that pertains to shikar, is full of hope, but it is not until 5.30 p.m.—much too late for such an affair—that we leave him perched in a machan chair for his first experience of sitting up for a tiger.

In the morning we find tracks of the tigress all around the kill, and learn she was heard sniffing under the machan early in the night but gave no other sign of her presence. She may have been within hearing when the machan was being made ready, or suspicious because the kill had perforce been dragged some twenty feet; or, more likely, had detected the occupant. Another all night vigil by our young friend afforded him the further experience of seeing five hyaenas at their loathly repast, for the weather was warm, and the poor buffalo hastening towards dissolution.

On the way back to camp we see tracks of other tigers, and a bear, along the sandy banks of the river, and at a village talk to an old man—eighty he says he is, perhaps he is sixty. A couple of months ago he heard a disturbance in the calf-shed near his hut. Entering with a man holding a lantern he found a large male panther had seized a calf by the throat. Graphically he illustrates how he severed the marauder's spine with one stroke of his axe and wholly disabled it with further blows, the final stroke almost cutting its head in two. Such killing of panthers with axes is not uncommon: even tigers have been slain in like manner by plucky herdsmen.

In the evening we take our first large mahseer of the trip, large for this river. He seized the spinning dead bait in a small eddy just off the main rapid along the further bank so was into the flood and eighty yards below in no time. The fish then reached more placid water at the entrance to a bay, but the clamber along the rocks took some time and there was a loose line now and then; a fall would have been painful, and dangerous in places. However all was well and the fish duly played out and netted. 30 lbs., 31 ins. long, girth 22 ins. a handsome mahseer with dark blue back and portly outline.

The sky is dark and lowering as we take the photograph. The *dhimar*, picking up his rod, hurriedly laid down when he heard our yell above the roaring of the waters, finds his minnow has been pouched by a mottled green snake and the line in a fine tangle. The light is now too bad for a photograph and we are eager for another fish. Several are seen to leap in the tumbled turmoil of the rapids; there is one savage tug at the bait but no hold taken; so we put up our tackle, trudge up the bank to the boat, and cross the river after the most gorgeous sunset it is possible to imagine.

This sudden activity of fish, and the flaming sky, was a sure presage of the storm which gathered next day to burst upon us from 3 p.m. to midnight. Every day we take coarse fish: *tengra*, *perrun*, *bekri*, eels, and realize, as indeed we well know, that this month of October is too uncertain for mahseer fishing in this river. March is the month. Then are golden days and silver nights, mahseer, tiger all delights! The days more red-hot than golden, but sport would compensate and risk of fever be less.

On the 25th October two small mahseer taken on spinning tackle are returned to grow larger, and the twin brother of our acquaintance of last Sunday is seen rolling like a porpoise in the swirling foam. Now the water is clearing, and there will be no more rain. Early in the month there were flights of duck and teal moving up and down, but none have been seen lately. Daily we see the three varieties of kingfisher, fish hawks, kites, ubiquitous crows, and blue rock pigeons. On only one day was a mugger seen and he, a monster of his kind, cruising down the flood like a Thames steamer.

When asked if any of the villagers are ever taken by muggers the *dhimar* relates a wonderful tale. 'It was,' he says, 'about the time of the great upheaval (meaning the Indian Mutiny) that a boy took two bullocks to the river to drink. A mugger seized the lad by the leg; he held on to the head ropes. The saurian would not let go, nor did the boy, but the bullocks were strong and dragged both boy and crocodile to the village where the latter was dispatched and the boy released'. A stout-hearted, also strong-armed village boy!

On the night of the 26th the tiger again took a stroll along the sandy shore of the river, passing between the two *bodas* without getting into touch with either of them. Yet the baits are tethered in carefully selected places where tigers have passed, and always do pass, on their nightly prowls. It is sheer bad luck and only two nights remain, for this outing must come to an end on the 30th October.

Now that the impetuous flow of the river has somewhat abated, the dhimar accedes to the suggestion that we can venture to the several falls and cascades accessible from the rocks in the centre of the rocky barrier. We go there in the afternoon and observe much of surpassing interest. Plain to the understanding are demonstrated the methods slow and sure by which the mighty force of the water continues to erode a way through the rocks and gradually destroy them. Very evident it is that during the ages to come the falls will recede further and further upstream.

First there is a small irregularity in the rock surface; a little sand whirled round and round forms a slight circular depression; then a few small pebbles continue the work; the boil becomes deeper, the pebbles larger; then arrive pebbles to size of a football and a larger 'well' is formed. All of these processes are seen at work in the various stages. The result is that two wells are formed in proximity to one another. As they grow larger the sides give way and an arch is formed. Gradually the arch is weakened at the summit and worn away; then the rocks are tumbled into the irresistible flood to be eventually ground into sand. So it is that the relentless power of Nature is ever at work and 'grinds exceeding small'. As the various processes of formation of these 'boils' and 'wells' can only take place during the few months of the year when the river is in flood, and then works intermittently and at various levels, ages indeed must have gone to the making of the great gorge that we see.

We ran to the principal cascade, where the two branches of it meet beneath an arch which will eventually collapse, and see two fat *mural* busily at work among the fry; and a large mottled-green snake—similar to he of the tangled line—is on a rock at the water's edge making quick lunges at the small fish leaping out of the stream as they essay ascent of the tumbled water. We see him catch one, and watch him dispose of it. Many he misses, and is not such a skilful fisherman as the 'pied fish-tiger' which hovers over a placid in-curving bay of the river further down.

At these centre cascades there is a great assembly of fish of all sizes, from minnows an inch long to fish of two or more pounds in weight: all of these in their countless millions are attempting the ascent, and adding the glitter of their silvery forms to the beauty of the foaming water.

The dhoti arrangement is soon at work to produce live bait, and in a few minutes a ten-pound mahseer, seizing it deep down under the arch, makes off down the further channel. The movement has been foreseen as, should the fish be able to turn the corner, all is lost for the line will be cut on the rocks and in any case the fish cannot be landed. So no line is allowed and the struggler hauled back by brute force, the hook-hold being fortunately sound, and directed into the nearer rapid. Once in the swirl of that he could not be denied and is fifty yards downstream in a moment; then he is played out and brought to net in a small bay close by.

A six-pound mahseer is caught, and two much heavier ones lost by the hook-hold not being good enough to withstand the drastic methods necessary to prevent them going down the further flood. As the day draws to a close a 14-lb. *perrun* is landed after a good fight in the heavier water of another fall, and then we reluctantly pack up to clamber over the rocks to the boat.

We have stayed overlong, and it is nearly dark when we commence to pole upstream. Soon it is realized that it is not possible either to see the rocks below the surface, or which way the currents set, or to pole against the current. To be swept down either of the main falls a descent of forty feet—as must happen if the boat gets into the heavy force of the stream—means certain disaster; so we perforce return to the rocks resigned to the necessity of remaining marooned for the rest of the night.

No food, shorts and shirt! But the sky is clear, the moon will be up in half an hour, and a cleft in the rocks is found which will afford some shelter from the chilly wind to be expected after midnight. The *dhimar* collects driftwood; so has a fire and his evil-smelling tobacco to comfort him.

Soon the familiar star patterns and constellations are picked clear upon a stainless sky shortly to be dimmed by the rising moon which sheds silver radiance over the upstream placid reaches of the river and enhances the magnificence of the wild turmoil of the falls and cascades. The night is very beautiful.

We wander about the rocks and watch the cascades and the leaping fish doubly silvered in the rays of the moon, a sight which will always live in the memory. The *dhimar* slaps his tummy to indicate it is time to roast some fish over the embers, so we are soon full-fed and ready for slumber which is somewhat slow to arrive owing to the vibration of the tortured rocks. The sun hat makes a sufficient pillow. The fisherman settles down close to the fire which he takes care to keep going through the hours.

Shortly after daybreak the friendly moon sets behind a giant mango tree on the further bank; the stars fade out in the pale blue sky, and the sun springs above the forest tree-tops

Pied fish-tiger (Pied Kingfisher)
Ceryle rudis

to shed welcome rays on the scene. The fisherman sets about making arrangements to get the boat away. A long rope is signalled for from camp. He wades and swims to the bank to fetch it and a man to help. At the head of the rapid the man is left near a rock, and the *dhimar* returns to the boat which we pole to a place near the edge of the difficult water. Then the rope is fetched by the otter-like native, and fastened to the boat; it is nervous work to watch him, hampered by the long rope, swept in an instant some twenty yards down before he obtains a footing.

The boat is taken as far upstream as can be managed, to whirl rapidly down when let loose, but be pulled with success to the eddy below the rock. At last the adventure is safely ended. As we arrive the swollen bulk of a defunct bullock comes floating down the river to be carried over one of the centre cataracts and be wholly disintegrated in the process. It is glad we are to have had a drink of river water before that came along!

In the afternoon we go again to the same place and have further success with mahseer and *perrun*, but take care to stop fishing in time to come away under similar arrangements to those of the morning. This is the end of the October outing. Six miles takes us to the forest rest house, near which is a sandstone fort about 150 years old in which European officials, ladies, children, refugees from Indore, were sheltered for some weeks during the troublous times of 1857, before they were moved for greater security to the Asirgarh Fort. The remaining eighteen miles to the railway station is done in a bullock cart during the cool hours of the night, and train taken at an early hour on the last day of October.

■ ■ ■

The Hunting Leopard
(*Acionyx jubata*)*

Surgeon Major T C Jerdon, (Indian Army)

Year of publication: 1874

T C Jerdon was born on 12th October 1811 and graduated as a medical student in 1829-30. He was appointed as an Assistant Surgeon in the East India Company in 1835 and reached Madras (Chennai) in February 1836. Apart from his medical duties his main interest was natural history. Jerdon could be termed a complete naturalist. From the time of his arrival in India he concentrated on studies on the wildlife of the country. His first publication was A CATALOGUE OF THE BIRDS OF THE PENINSULA OF INDIA published in 1839-40. His Catalogue included 420 species. He then published his 'Illustrations of Indian Ornithology' in two volumes between 1843 and 1847. Jerdon's proposal for publication of a series of monographs on Indian Zoology was accepted by the Government of India. His volume on ornithology describes 1,008 species, his volume on mammals describes 247 species, but he was unable to complete his volume on reptiles and fishes owing to ill health. Jerdon's mammal and bird volumes were the points of reference for the many studies on the mammals and birds of the subcontinent in the 19th century. He was also a botanist of repute. He died on 12th June 1872.

Common Names: *Chita*, Hindi—*Yuz*, of the trainers.—*Kendua bagh*, Beng.—*Laggar*, in some parts.—*Chita puli*, Tel.—*Chircha* and *Sìungi*, Can.— Eng. *Cheeta*, or Hunting Leopard.

Descr.: Bright rufous-fawn with numerous black spots, not in rosettes; a black streak from the corner of each eye down the face; tail with black spots and the tip black; ears short and round; tail long, much compressed towards the end; hair of belly long and shaggy, and with a considerable mane; pupils circular; points of the claws always visible; the figure slender, small in the loins like a greyhound; limbs long.

Length, head and body, about 4½ feet; tail 2½ feet; height 2½ to 2¾ feet.

This animal was the original *Panther* and *Leopardus* of the ancients, who considered (with the Arabs of the present day in Northern Africa) that it was a breed between the lion and the pard.

The hunting leopard is found throughout Central and part of Southern India, and in the North-west from Kandeish, through Sindh and Rajpootana to the Punjab.

It is also found in South-western Asia, as far as Syria and Mesopotamia, and throughout Africa. It is stated to exist in Ceylon (fid. Baker ex Blyth), but I doubt extremely its occurring in that island. I have met with it myself in the Deccan, near Jaulna, and near Saugor in Central India, in both cases in tolerably open ground where the common antelope was abundant. In the one instance I turned it out of a small low *bér* bush, along with a jackal that was keeping it company; and near Saugor I saw a pair of them stalking some nil-ghai in mid-day. I had one young one brought to me also at Saugor, only a very few days old. It was clad with long hair of a greenish fawn-colour without spots, and it was not for several days that I recognized it to be the Cheeta:—the cheekstripe was the first mark that appeared. Antelope, gazelle, and nil-ghai are said to be its chief food in the wild state, but it is said occasionally to carry sheep off. Native shikarees assert that it usually has its lair among rocks, and feeds only every third day, sleeping the two others.

I brought up the young one above alluded to along with some greyhound pups, and they soon became excellent friends. Even when nearly full grown it would play with the dogs (who did not relish his bounding at them), and was always sportive and frolicsome. It got much attached to me, at once recognizing his name (Billy), and he would follow me on horseback like a dog, every now and then sitting down for a few seconds, and then

*Current name *Acinonyx jubatus*   I^st edition 1867; IInd edition 1874

The genus *Cuon*, to which the Asiatic Wild Dog (*Cuon dukhunensis*) belongs, is distinguished from the genus *Canis* by the more rounded ears and proportionately shorter muzzle; by the line of the face viewed sideways being slightly convex, that of *Canis* being straight or concave, and by having only two true molars on each side of the lower jaw instead of three. The mammae are more numerous, there being usually 6 or 7 pairs instead of 5 typical in *Canis*. Blandford mentions the long hair between the foot-pads, but this is not unusual in many species of *Canis*. In other respects *Cuon* agrees generally with *Canis*.

SPECIES AND RACES OF THE WILD DOG

The Asiatic Wild Dog has a wide habitat. Pocock (P.Z.S, 1936, p. 34) gives its distribution as ranging from "Saghalien, Amurland and the Altai Mountains, about lat. 50° N., over the whole of continental Asia, roughly east of Long. 70° E., and occurring in the Islands of Sumatra and Java, but not in Japan, Ceylon or Borneo'. Within this wide range the older Zoologists claimed to recognise two or even three distinct species. Blanford recognised a northern and central Asiatic species under the name *C. alpinus*, an Indian species, *C. dukhunensis* and a Malayan species which he called *C. rutilans*. Pocock (loc. cit.) having at his disposal the great mass of material made available in recent years, concludes that in all Asia there is a single species of Wild Dog, to which he assigns the name *Cuon javanicus*; *javanicus* is the name first applied to any form of *Cuon*, and it therefore antedates *dukhunensis* by which our Wild Dog has become more generally known. Pocock recognises a number of local races. The form found in the Indian Peninsula, south of the Ganges, and also in Assam, he names *Cuon javanicus dukhunensis* Sykes, distinguishing it by its larger skull from the typical Javan Wild Dog (*C. j. javanicus*) and from *C. j. sumatrensis*, found in Sumatra and the Malay Peninsula, by its longer, fuller coat, yellower, less vividly red-colouring and again by its larger skull. The race found in the Central and Eastern Himalayas through Kumaon, Nepal and Sikkim is *Cuon j. primaevus*. Pocock distinguishes this race from the Peninsular form as being on the average redder in hue, fuller in winter coat, more amply provided with under wool and more hairy-soled. These distinctive characters he states are evident from puppyhood. The Wild Dog of Kashmir is recognised as a separate race to which, the name *C. j. laniger* is assigned. Pocock distinguishes the Kashmir from the Central and Eastern Himalayan Wild Dog by its much fuller, softer coat and much paler colouring. No material is yet available to establish the identity of the Wild Dogs of Burma, though Pocock has described a new race from Moulmein for which the distribution given is 'North Tenasserim and possibly Annam'. To this form he gives the name *C. j. infuscus*, distinguishing it from the *javanicus* and *sumatrensis* mainly by cranial characters.

AJAY DESAI

THE INDIAN WILD DOG

Coloration

From time to time sportsmen bring to notice variations in colour and in size. These can safely be said to be caused by environment, more or less abundant food supply, and climatic differences in various parts of the extensive region inhabited by this widely distributed animal.

Blandford describes the colour of *C. j. dukhunensis* as follows:

'On upper parts generally rusty-red varying in some specimens to rufous grey or even light brownish grey, paler below. The colour is generally not uniform, being variegated by dark tips to the dorsal hair. The under-fur, when present, varies in colour from light brown to dull rufous on the upper parts, and has light coloured coarser hairs intermixed; the longer hairs are light rufous, with dark rusty-red tips. The terminal portions of tail black (very rarely the extreme end is whitish). The young animals are sooty brown throughout.'

Dunbar Brander's description is:

'Uniform red, shading into yellow or dirty white on the belly. The points of the hairs along the dorsal ridge are often black. The ears, which are pricked, are frequently pointed black. The tail, which is short (about 8 inches) has a bushy tuft of black hair some 5 in. or 6 in. long at the end. There are generally a few grey hairs in the middle of this black tuft, and it is not uncommon for these hairs to be sufficiently numerous to amount to a small white tuft within the larger black tuft.'

He adds, differing from Blandford as far as the wild dog of Central India is concerned, that the existence of the white tip is much more common than the black tip.

In an Editor's foot-note at page 516 of Vol. XXXI of *JBNHS* it is remarked:

'In colouring the wild dog varies from uniform red to rufous grey or even light brownish grey. As regards the colouration of the terminal portion of the tail,—of a series of thirty-four skins obtained in India and Burma all except four have black tips to the tail, including specimens from South India and Canara—of ten specimens in the Society's collection obtained in the Berars and C.P. seven have black tips and three white.'

There are often white hairs at the end of the tail not visible unless the longer black hairs are parted. It is recorded that three three-quarter grown specimens shot from the same pack (U.P.) had: one a black tail tip, one a few white hairs, and the third a distinct white tip to the tail like that of the Silver Fox.

Mr. R.C. Morris notes that wild dogs, when killed, seem to lose immediately the 'gloss' of their coats: so also with Hyaenas.

Recorded weights and measurements of *Cuon* to be found in the *Journal* and other publications are tabulated below:

Vol. VII p. 503.	Head and body 34 in.	Tail with hair 17 in.	Height 20 in.
Vol. X p. 449.	Head and body 35 in.	Tail with hair 17 in.	Height 19½ in.
Vol. XIII p. 529.	Head and body 34½ in.	Tail with hair 17 in.	Height 21½ in.
Vol. XIII p. 529	Head and body 34½ in.	Tail with hair 17 in.	Height 18½ in.

All the above are females.

Vol. XXXII p. 714.	Head and body 38 in.	Tail 14 in.	Weight 28 to 32 lbs.

This is from Kashmir and gives average measurements.

Dunbar Brander: M…. 22 in. at shoulder, weight 43 lbs.
                 F….. half an inch shorter and 5 lbs. lighter.
It is apparent much more data are needed.

Breeding, Taming, Characteristics, and Disposition

The mammae may number as many as sixteen. Most writers say fourteen. Eight pairs are mentioned by Pocock. The mammae are not necessarily in even numbers on either side. Eight on one side and six on the other side have been counted.

The number of young at a birth may vary from two to six or more. Litters of seven and ten have been seen by me, also a case of seven embryos being taken from a shot animal. Major Phythian-Adams tells me that he himself took nine embryos from a wild dog shot by him on the Nilgiri Hills; and his chauffeur took seven pups from an earth where there was a bitch from the pack on guard. At this place a number of the dogs were breeding; there were many earths; a regular pack nursery! Such places are not uncommon: I recollect one in the Biba shooting block, C.P., where there were several earths–crevices in rocks in a nala bed.

'Robin Hood', in Vol. X, page 127 of the *Journal* described his six puppies as 'six lovely little russet-red balls of fur'.

They were reared by a village pie. The first pie, which had five pups of her own, was so horrified at their odour that she would have none of them. The second pie, with three pups, after a great deal of trouble, became as attached to the jungle puppies as to her own. This was no doubt due to the fact that, in the course of a week or so, her own puppies got to smell as badly as their jungle confreres. At a very early age the jungle puppies evinced an incorrigibly pugnacious disposition, and fought with appalling ferocity. It was incredible to see such small things fighting with so much resolution and tenacity. If the uppermost belligerent were lifted by the tail to the height of one's head it would carry its opponent up with it, and the latter would likewise decline to relinquish its hold. They would fight in this way day after day, the sluts being quite as tenacious as the dog puppies. The unfortunate pie-pups must have wondered into what company they

A.J.T. JOHNSINGH

had fallen. They would yell piteously when the wild foster brethren shook them up, and we were obliged to send them away when quite small to prevent them getting killed. Meanwhile the wild pups continued to fight with each other with unabated ferocity until they were about seven or eight months old when—most singular to relate, they put a permanent period to their hostilities and lived in perfect amity. Apparently they had decided which was the strongest amongst them, as they paid marked deference to one large dog–the largest among them—who acted as their leader. They never fought with each other after they became adults. The big dog above referred to had a white spot on the near forepaw and the extreme end of his tail was tipped with white. This was observable only on a close and critical scrutiny. Two other dogs were similarly marked; the remaining three were entirely russet-red. They all had large prick ears (which they laid back flat like a vicious horse when angry or attacking), and long, heavy, bushy tails. They would eat nothing but raw meat. We nearly starved them to death in the endeavour to make them eat cooked food, but without avail. They would eat nothing but flesh, and not that unless it was raw. They would not eat stale meat…. The dogs were dangerous to approach when feeding; but could be handled at any other time.

They would never molest men; but would 'go' for any and every animal. In consequence they had to be kept on the chain. When they broke loose, which not infrequently happened, they did not attempt to escape, but always kept about the house. They all met untimely deaths. One died while still a pup from confinement in a basket—they appear to require plenty of fresh air—another died while en route to a railway station; and the remaining four from diarrhoea engendered by a cannibal propensity they had of killing and eating any stray dog they could get hold of.

My observation of these dogs has convinced me that for gameness, staunchness, and invincible tenacity we have no breed of domestic dog to compare with them.

Major Phythian-Adams tells me that the above exactly corresponds with his own observations of wild dog puppies kept by him.

The wild dog can be tamed, but has to be taken in hand when very young. A pair possessed by Mr. Charles Theobald of Mysore for about two years were not related, having been captured in the forest in different localities, and were about a month old when he got them. At first they had the strong smell characteristic of most wild dogs, but this wore away with regular baths, and then they smelt the same as domestic dogs. Both matured in about a year, and were mated. One pup was born in November but was accidentally bitten by the mother at time of birth, and died.

Although the dogs were quite tame they were difficult to manage and were not savage. No strangers could go near them. The period of gestation was about the same as that of the domestic dog. The noise the dogs made most was the usual soft whistling one to call to each other, or to me. They would come to me when called. When distressed they would whine, and utter a suppressed bark when angry.

When excited they wet themselves, and the urine would be splashed about by whisking tails. The urine had no bad effect on human eyes. That was a personal experience.

They were fed with milk in the mornings, raw meat in the day, and ordinary dog's food in the evenings. Sheep's leg with hair on was regularly supplied. Food was digested and passed in about twenty-four hours.

As to the last sentence it is probable that, in a wild state, food is passed in a shorter time than this. Phythian-Adam's pups were fed much as above.

It will be noticed that this feeding experience differs from that of 'Robin Hood'. It is a pity that the gestation period was not exactly noted, as there appears to be nowhere any record of this.

It may be noted here that until the young ones are able to leave the place where born they are fed by partially digested meat vomited by the mother. Breeding season is from November to February.

Popular Names

To the naturalist and the general public *Cuon* will always be known as 'Wild Dog' although, as Blandford says, the name is clearly a misnomer; for in every important detail in which the genus *Cuon* differs from *Canis*—form of skull, dentition, number of mammae—domestic dogs agree with the latter and not with the former.

Vernacular Names

Assamese: *Kuang-kukur* or *Rang-kukur*, **Bengali**: *Ban Kutta* or *Ban-kukur*, **Bhutanese**: *Phara*, **Burmese**: *Tan-kwe*, **Canarese**: *Ken-nai, chen-nai*, **Chenchu**: *Reis-kuku*, **Chin**: *Nyar*, **Gond**: *Nerka*, **Gujrati**: *Eram-naiko*, **Ghurkali**: *Ban-kukur*, **Hindi**: *Adivi-kuta, Son-kuta, Sona-kuta, Rasa-kuta*, **Hindustani**: *Jungli-kuta, Ran-kuta, Ban-kuta*, **Kachin**: *Kyi-kwa-lam*, **Kashmiri**: *Jungli-kuta, Ram-hun, Ban-kuta, Bhansa* (E. Himalaya), **Korku**: *Bun-seeta*, **Lepcha**: *Sa-tun*, **Mahratti**: *Hahmasai-kuta, Kolsun, Kolsa, Kolasri*, **Malayalam**: *Shen-nai*, **Nepali**: *Bwaso*, **Malay**: *Sirgala, Arjing-kutar*, **Tibetan**: *Hazi, Phara*, **Malayalam**, **Tamil**: *Chen-nai*, **Telugu**: *Vanna-koota*.

Odour

Blandford remarks that the strong and unpleasant odour of *Cuon* resembles that of the jackal, this being due, in part, to secretion from the anal glands. But this is not always the case, as some observers have described the wild dog as having the exact smell of the

domestic dog; e.g., Inverarity, who was an experienced and reliable sportman, who relates having shot one of a pack, 'it was a female and had the exact smell, of a domestic dog'.

Voice

The difficulty in describing the various voice sounds of the wild dog is apparent from the writings of naturalists and sportsmen. The word *bark* has been used but does not convey a correct impression. Voice noises *growl, snarl, whine*, are almost the same as those of the domestic dog. The mate call mentioned by Dunbar Brander I have, unfortunately, not heard. The sound uttered when startled, alarmed, or at time of disputing a tiger or panther kill, as I have myself heard, is a sort of hyaena-like chattering—'analogous to the 'chuck, chuck, chuck' of the Indian fox.' (Dunbar Brander).

The sound used when the dogs are communicating to one another otherwise than when hunting is difficult to express. A sort of soft whistling noise seems best to describe it. 'Their usual call is a highly pitched whine'. (Best).

Other noises described are: 'A weird bewildering noise—such as I had never before heard in the jungles–' described afterwards by a friend who was with him at the time as 'a kind of fiendish hysterical yapping, in a shrill chorus, decidedly uncanny and all-pervading'. This was when Professor Littledale's terriers rushed into bamboo cover near his camp in the Central Provinces. The hunting notes are variously described. 'Hawkeye' mentions 'a tremulous whimper'. Havelock 'a whistling howl when running to view'. Inverarity saw two at a shallow stream, halt at the water's edge and utter 'loud wailing howls'; these, until they reached the water, and on another occasion a pack also ran mute. I have seen dogs running mute several times and think that when first on the trail they run mute, when the scent tells them they are nearing the quarry they whimper, and when they run to view, or are closing on the quarry, they break into what has been called

KEDAR BHIDE

'full cry': this is not the music of foxhounds but 'an indescribable howl'. The above is not always the case. The hunting cries have also been described as a 'sort of yapping bark'; this in Nepal. Major Manners Smith could not say they were 'in full cry' but they were evidently hunting, probably *thar* or barking deer. In Burma Mr. C.E. Milner one night in Tharrawaddy heard a pack 'in full cry, rather like a poor-voiced pack of hounds.'

Whistle, whine, whimper, yap, are *voice noises* mentioned by Major Phythian-Adams, who has never heard them 'bark', as they are said to do by 'Hawkeye'. Best also uses this word, but it is not the bark of a domestic dog: a more hoarse sound. Colonel A.E. Ward writes 'when running by scent they only whimper, but when the prey is in sight and at hand they often break into an indescribable howl'. A night hunt in Nimar is described as 'frenzied whimpering cry'. Mr. La Personne writes of wild dogs hunting at night as 'baying' but that would not, I think, convey a correct impression.

The Indian Wild Dog does not *bark*: domestic dogs turned loose on the island of Juan Fernandez quite lost their bark after 33 years. (Sterndale).

Calling up

Major Phythian-Adams informs me that wild dogs can be successfully called up by a series of 3 'toots' on an empty 318 cartridge case, these representing the three whistling notes of the dogs calling to one another. This method is best employed after one of a number has been shot; but will also call up from nearby cover a single dog, or one of a pair, which has been seen. Another method found successful is the use of a leaf blown upon edgeways between the thumbs, as employed in Burma for barking deer. It is useful when no shot has been fired. This leaf noise has been mentioned by writers in the *Journal*, and is known to many sportsmen. It is mentioned by Dunbar Brander.

Best (INDIAN SHIKAR NOTES) describes a similar leaf noise used by him with great success. R.C. Morris describes their calls to each other as a shrill 'ow, ow, ow', the noise being similar to that produced by blowing into a medium bore cartridge case, as described by Phythian-Adams.

Blinding Eyes of Animals with Urine

Most writers remark upon the belief held by jungle inhabitants of all parts of India where *Cuon* is found that the animal deliberately makes use of its urine to blind game being hunted, either by sprinkling bushes with it and then driving the animals through them, or whisking it into the animal's eyes with their tails. This belief, which will probably never be relinquished by the jungle inhabitants of India and other Eastern countries, is mentioned by Dr. John Fryer in very early days; also in Williamson's 'Oriental Field Sports'. Blandford remarks that a somewhat similar belief is held as to wolves in parts of Europe.

Although we know much of wild life lore of jungle peoples is accurate, or has a substratum of fact, yet in this matter it can be safely asserted that the wild dog's urine has no special blinding property, and that the animal does not deliberately use its urine as an aid to hunting.

Likes and Dislikes as to Meat

In his GAME-BOOK FOR BURMA Mr. Peacock rightly states that wild dogs generally demolish their kill at one sitting, but adds that they desert their kills, if any meat is left, at the first sign of decay: also that he has not known or heard of their touching carrion, though it is likely enough they may do so when very hungry. Dunbar Brander remarks that high or contaminated meat is distasteful to them and their kills are therefore nearly always finished off by vultures.

It is within my own experience on several occasions, and that of other sportsmen, that wild dogs will appropriate a tiger or panther kill should they come across it. No doubt they prefer fresh meat—that of animals they have themselves hunted, but they do not refuse meals accidentally found. In the C.P. a large pack demolished the carcase of a skinned tigress thrown outside my camp. R.C. Morris notes that he has personal experience of at least a dozen stale kills being eaten by wild dogs in the Billigirirangan Hills of South India, and the North Coimbatore forests adjacent to them.

The following note shows that they will sometimes eat very foul and disgusting meat, but it is perhaps a rather exceptional case. R.C. Morris writes:

'I recently shot two wild dogs, a male and a bitch, which were as badly affected with mange as I have seen on any pie-dog. The dogs were feeding ravenously on a three-day-old tiger kill and seemed to be quite active. With the exception of the head, which was blotchy, and the ridge of the back the dogs were almost devoid of hair including the tail. The short pig-like tail, free of the brush or hair, and the pink and mangy flesh appearance of the flanks and stomach gave the dogs quite a revolting appearance.'

Peacock also writes that he has never known his tiger or panther baits to be found or killed by wild dogs. This is contrary to the experience of sportsmen in India, more particularly in the Central Provinces, where it is not uncommon occurrence for tethered baits to be killed by wild dogs.

Domestic Stock

Wild dogs do not ordinarily attack herds of cattle grazing in the forests, though there have been a few instances of the kind recorded. The Author of Nilgiri Sporting Reminiscences' (1880) writes that:–'…they make frequent attacks upon the Badaga's buffalo calves, and the ryot's sheep and cattle of the low countries. A pack of fourteen or fifteen wild dogs, about five years ago, committed a raid upon a herd of about thirty calves whilst out grazing, belonging to the Westbury Estate, Segoor, and killed five and wounded two of them which died some days after; before I could get with my guns to the spot they had been scared away.'

Such an affair as the above is exceptional, but Phythian-Adams tells me that he knows several cases of buffalo calves being killed out of Badaga herds near Anaikatti: and R.C. Morris notes to me that under certain conditions, in the absence of game animals, it is within his experience that wild dogs will stay in a locality and kill cattle. They did not kill cattle all the time, but did so in the dry weather when the sambhur were noticeably scarce. He adds that some years ago there were a large number of pig in those hills (Billigirirangans): wild dogs then increased and killed scores of pig; since then pigs have become quite scarce.

In the other parts of India they occasionally kill single cattle and goats, but it is not a common occurrence. In Kashmir they often worry sheep in the open country'.

R.C. Morris sends me the following interesting note:-

'Wild dog, a large pack, killed 25 calves here in one year, including six in a go, that were grazing together. This occurred at the time when wild dog were numerous. Stray mature cattle were also killed and devoured; so much so that the Badaga herdsmen at Bellaji (where you camped once) asked me for poison. I gave them 'Atlas' with instructions to dilute with two parts of water. This the herdsmen did not do, but used the neat poison on a freshly killed carcase, and then retired to watch events. The dogs returned and completed their meal, but were soon lying around in great distress. The Badagas told me that there were about twenty in the pack, and as the reward for wild dog is Rs. 10 per head in Kollegal and North Coimbatore they

were delighted at the prospect of earning Rs. 200 by the poisoning of the whole pack. To their consternation however the dogs, one by one, commenced to vomit up what they had eaten, and eventually made off. Before this they could have been clubbed to death easily, apparently, but the Badagas were afraid to approach the pack. I was away at the time. This had the effect of curing the dogs' desire for beef for some time.'

It is curious that wild dogs do not molest bullocks in a cart, or attack a pony ridden past a pack lying about by the side of a forest road. Postal runners, frequently killed by tiger or panther in certain parts of India, are never killed by wild dogs. It is fortunate indeed that they have a non-hostile disposition to man and, mostly, to domestic animals.

Game Animals

Wild dogs kill a great many game animals, as also wild pig. To their diminishing the stock of the latter there is no objection, for these do much damage to cultivation; but their ravages among game animals are often very serious. With the exception of fairly mature bison and buffalo they kill almost all forest animals.

It is fortunate that they do not kill, as do the Dingo of Australia, for the sake of killing, but only when hungry. An instance illustrating this is noted to me by Mr. R.C. Morris who, with Mr. C.W.G. Morris, saw a pack of wild dogs, obviously well-fed, lying about on a grass hill-top while sambhur walked, tails stiffly erect, right up to them. Apart from eyeing the deer lazily the dogs did not stir. After a few minutes the sambhur trotted off for a short distance and then commenced to graze! Some time later the dogs moved off slowly in the opposite direction.

On the Nilgiris Plateau the wild dogs kill Nilgiri Langur which probably leave the safety of trees and race across the open slopes. (Phythian-Adams).

At page 165, Vol. XLI, Mr. Dunbar Brander comments upon the extraordinary behaviour of monkeys when attacked by leopards and by dogs (presumably domestic dogs), having known them to abandon the safety of the tree-tops and take to earth to their destruction, and remarks 'In this respect an animal which must be considered as intelligent, behaves like an imbecile.'

I, also, have observed similar instances, but have noticed that it was seeing the human being the monkeys, or lungoors, vacated their arboreal safety; and the same has been the experience of friends with whom I am staying near the Periyar Lake, Travancore, at time of writing. I have known, on a moonlight night, a pair of panthers acting in concert chase lungoors out of a tree: this being no doubt a normal method of hunting these creatures to their destruction.

That the imbecility remarked upon by Mr. Dunbar Brander is not confined to the lower species of Primates is instanced by the fact that a woman of the Kumaon Hills— sister of the village shikari employed by me in 1924 during pursuit of a man-eating tiger—was killed by the tiger because, when in the safety of a tree gathering oak leaves for cattle fodder, other women being in adjacent trees, and the tiger came under her tree, she cried out 'The tiger has got me! The tiger has got me!' and fell from her secure perch into the animal's jaws!

A week or so previously her aunt had been taken by this tiger when similarly gathering leaves, but was stalked while on the ground.

It would seem that Nilgiri Langurs of the Nilgir sholas are frightened out of trees by wild dogs, and perhaps the same thing happens in forests of the plains, but I rather doubt it.

Mange

That many pariah dogs are mangy we know, and this may be communicated to wild dogs. But wild dogs do not ordinarily have much to do with village dogs. In addition to the instance above quoted Mr. Morris informs me that when some years ago wild dogs were at their peak in numbers in the Billigirirangan Hills they appeared to have developed mange very badly. In 1938 three very mangy wild dogs galloped down the ghat road for some distance in front of his motor car.

Major Phythian-Adams draws my attention to the fact that mangy skins are sometimes produced for rewards in the Nilgiri Hills and writes: 'Some of the dogs killed in 1937 were in a very mangy condition; and a number were found dead, in 1893/94, probably from distemper, sometimes as many as three and four together in one spot.'

Besides the above instances from South India there is a case of mange noted from the C.P. at page 1046 of Vol. XXIX.

It is likely that wild dogs, living in burrows and holes in the ground and among rocks as they do, may develop their own skin diseases.

Attitude to Mankind

I have been able to find only two instances of wild dogs being said to have been aggressive to man. Colonel Caton Jones (Vol. XVIII, page 194) relates that the wild dogs of Nimar were very bold; that they growled at him several times, and that just before he left the jungles the Forest Ranger informed him that four or five of them had attacked two forest guards who had killed one dog with an axe. He (Col. Jones) feared that unless these wild dogs of Nimar were killed they would soon become man-eaters. That was in 1907, but there were no subsequent happenings.

'Robin Hood' relates in his article—Vol. X, page 127:-

'As I was walking along a game-track in the Nullamalais I came upon a dog stretched across the path. Instead of bolting away, as wild dogs generally do at the sight of man– the dog rose up reluctantly and slouched in a semi-circle, eyeing me with a sinister look. I was unarmed. The dog at length disappeared behind a bush, and I walked on marvelling at its strange behaviour. I had gone thus about a furlong when I happened to look behind and saw the dog rushing after me at full speed, with its nose to the ground (this was strange as it had already seen me). I immediately faced round with a large stone which I hastily picked from the ground. The dog rushed almost to my feet (still with its nose to the ground and not looking up!) and I hit it a severe blow with the stone, at the same time rushing to meet it with a loud shout. The shout appeared to alarm the dog more than the missile. It started aside and again semi-circled–while I retreated backwards-keeping my face to the dog in the direction I had come. I did this as I should have gone into thicker jungle had I gone on. I felt convinced that the dog would again attack me there, and perhaps fetch other dogs to its assistance. In this way I backed out of the jungle to my tent. It was fortunate for me that this dog was alone. Had there been others with it, its aggressive demeanour would probably have incited them to attack.'

He sent for his gun, which had gone by a road to meet him further, on, and remembering the dog was a slut with dugs nearly touching the ground searched the forest and eventually found a cave concealed in the undergrowth from which the six puppies were taken. During the night the mother came round the camp and the servants were throwing fire-sticks at her all night to keep her off. My impression as to this incident is that there was no intention to attack; it was the very natural desire of a mother to see a possible aggressor off the premises.

Native shikaris of Anaikatti below the Northern slopes of the Nilgiris say that if a single man comes on a pack eating a kill they will stand their ground: also that a mother bringing meat to her puppies' earth will demonstrate.

European sportsmen have not recorded such experiences; but the note of Caton Jones above cited may be referred to.

It is fortunate that the attitude of the wild dog to human beings is almost invariably wholly unaggressive: had it been otherwise, mankind in the forest areas would never have been safe from their attacks.

Altitude to Larger Carnivores

No writer on the wild dog of India omits discussion of the widely prevalent belief and assertion of native shikaris and jungle people that wild dogs will, on occasion, attack and kill the tiger; Jerdon, Sterndale, Baldwin, Sanderson, Littledale, Inverarity, Dunbar Brander, Best, Peacock, and others, all touch on this subject.

Dunbar Brander gives two pages of his book to discussion of the question. He once witnessed wild dogs annoying a tiger: and relates a very circumstantial account by villagers who heard a fight in progress, and when after some time they timidly approached the spot found a dead tigress and two dead wild dogs. He concludes that there can be no reasonable doubt that they do occasionally kill tigers.

In his article in Vol. X 'Robin Hood', evidently a Forest Officer of South India, relates occurrences in the Nullamallai Hills, east of Kurnool and south of the Kistna river, where Chenchus are the aboriginal tribe, which converted him from his previous scepticism.

Perusal of 'Robin Hood's' article takes my memory back to 1902 and the Bairnuti Inspection Shed near which Robin Hood witnessed the killing of a Chenchu she-buffalo by a tigress which killed her calf and then, the mother defending her baby, the slaying of the marauder by the angered herd. The heat of those jungles. Terrific!

At page 218 of his very interesting book *Leaves from the Diary of a Soldier and Sportsman*, Major-General M.G. Gerard, who, as he himself told me, had shot 227 tigers before he left India, writes:-

'Two sowars of my Regiment (The Central India Horse), who had been out prospecting for me, brought back some scraps of tiger skin as big as a napkin which they had found under the following circumstances. They were informed at one jungle village that a few days previously a tiger had been seen on top of a rock on the plateau above, surrounded all day by a pack of wild dogs. During the night they heard 'a tamasha' as they termed it, and upon my men, accompanied by some of the villagers, repairing to the spot they found the scraps of skin above mentioned.'

I have recently (January 1940) obtained an authentic instance of a full grown tiger having been killed by wild dogs. Mr. G.A. Tippets-Aylmer, a Planter in the Wynaad, South India, tells me that one day some years ago, when in the forests surrounding his Estate, he came across the skeletons of a full grown tiger and seven or eight wild dogs. These were, perhaps, ten to fifteen days old and had been eaten and pulled about by jackals, pigs, porcupines etc. By the bushes having been laid flat and other signs it was apparent that a great fight had taken place. Perhaps, in this instance also, the cause of the combat was interference by the tiger in a hunt in which the pack was engaged.

It has been sufficiently established that wild dogs of the Indian jungles can, and do, kill tigers. Such happenings may not be very common. No case of the kind has come to my personal notice during jungle excursions in many parts of India scattered over a period of fifty years in this country.

It may be that some of these attacks are the result of a quarrel, or for the sport of baiting these animals (Dunbar Brander), but probably also for food when they have not encountered other game; or on finding the blood trail of a wounded beast ('Robin Hood'); and, more likely than all, on the killing by a tiger of an animal being hunted by the dogs and the ensuing fight for their quarry by the ravening pack. That they also kill panther and bear is shown below.

There are several recorded instances of wild dogs attacking panthers and bears. In Vol. V, at page 191 is a case of a large pack of ten or twelve couples treeing panthers in the Central Provinces. Had the business not been interrupted it is likely one or both of the panthers would have been killed. At page 194 of Vol. xviii wild dogs are said to have treed a female panther; and at page 218 of Vol. XXX, R.C. Morris describes having seen wild dogs attacking a bear which they would have killed had not a companion's rifle intervened. The dogs were quite mute during the fight. He also relates that he once saw the remains of a panther killed by wild dogs in a patch of Sholaga cultivation. The jungle people (Sholagas) described the fight to him. On that occasion no dogs were killed by the panther: but that the reverse takes place was the experience of another sportsman of those same Billigirirangan Hills who found the remains of two wild dogs which had been killed and eaten by a panther. In the same article is the account given to him by his tracker of a tiger having been killed by wild dogs. At page 744 of Vol. XXXVI is a note of a panther having been torn to pieces by wild dogs. It seemed that the panther had pounced upon one of the dogs which went to drink at a pool and the noise it made had brought the pack from the neighbouring cover. No trace of any other dog being killed in the fight was found. Following this note is an account by Colonel J. Pottinger of having witnessed a panther being driven off its kill at 9 p.m. on a bright moonlight night by a pack of ten or fifteen wild dogs.

Mr. L.E.C. Hurst tells me that he saw a bear treed by wild dogs in the Chanda forests, C.P.

No instance of hyaenas or jackals being killed by wild dogs is related. Perhaps there is, as Dunbar Brander remarks, a blood-brotherhood bar.

Wild Dog versus Domestic Dog

A case of a wild dog being killed by two domestic dogs—a Ceylon Beagle and a Cross-bred Airdale-cum-Irish terrier—is recorded at page 949 of Vol. XXXVII. The wild dog was three-quarters grown and alone among the tea bushes. At page 343 of Vol. XXIII is a case of two pariah dogs keeping a panther off his kill after dark. And at page 428 of Vol. XXXIII of a pariah dog tethered for a panther near a village, successfully resisting the effort of the pard to make a meal of him until the watcher in the machan was able to plant a bullet. All pariahs are not curs.

At page 200 of Vol. XXXIII Mr. Peacock asks for an instance of a single domestic dog successfully joining battle with a wild one, and gives a photograph of his bull-terrier who would have probably given the Malay wild dog pictured with him 'a very thin time' had they clashed teeth!

'My dog chased a wild dog but others came and followed him to within a few yards of me' (Pythian-Adams). Dunbar Brander mentiones a similar incident actuated, probably, by curiosity.

Methods of Hunting and Seizing

The wild dog mostly hunts by day, especially in the early hours, but not infrequently on moonlight nights, and occasionally on dark nights. They are on the move early in the morning, and it is then they get on the trail of some animal the fate of which is sealed

once the hunt has begun in earnest and they have 'settled down to the long, lobbing canter that can at the last run down anything that runs.' That *sometimes* packs will relinquish the chase is instanced by a stag sambhur, hunted to a river at a place where there was a long, wide pool swimming to the further bank, the pack having him in full view. Yet the dogs abandoned the hunt. That was on the Denwa river, C.P. (L.E.C. Hurst).

They must have excellent noses. Littledale's *Cuon* bitch sought her food more by scent than by sight. Often when she did not see clearly where a bit of meat had fallen she would nose it out with great quickness. In this respect her sight improved: at first it was very bad. He describes the characteristics of the wild dog to be:- 'fierce yet shy; no amount of training could teach it to be gentle; shyness and distrust of man; fierceness and currishness combined; swiftness in snatching; tenacity in hanging on these are the strong points.'

'In the adult animal' he observes, 'the senses of hearing, sight, smell, must be developed to an extraordinary degree of perfection, judging from this animal alone.' That this power of scent is extraordinary is shown by the fact that they can follow a line in the hot weather when the ground is dry as a bone.

Besides being provided with all the highly developed senses necessary for successful hunting, as also extraordinarily muscular bodies, the wild dog has acquired, as has the wolf, an aptitude for team work. That they hunt in concert is certain. This has been seen in the Nilgiri hills where open hillsides, wooded hollows, and distant views provide ideal conditions for such observations.

'Around Anaikatti and Mudumalai at the foot of the Northern slopes of the Nilgiris wild dogs come out of the jungle on to a forest road about 7 a.m., and there idle about for some time performing their morning offices and, presumably, discussing the day's plans. They must be able to communicate or team work would not be possible.' (Phythian-Adams). Those who have kept dogs of various breeds must have noticed that they undoubtedly communicate ideas.

In *A Game-Book for Burma* the author remarks that he has never seen a wild dog's kill at any distance from water; and that the quarry is almost invariably run into, or runs into, a pool or a stream of water in which it is bayed or torn to pieces.

Hunted deer undoubtedly take to water, when there is any, wherein they have some chance, if not too submerged, the stags by use of antlers and those without by striking with fore-feet, of defending themselves. But there are not many pools of water, and no streams, in many dry jungles where wild dogs are numerous; so it is probable that the main cause of hunted animals being killed in pools and streams is not that the water is a place of refuge, but because hunted animals are naturally forced more and more down-hill as they become exhausted. Most parts of Burma are better watered than much of the jungle country of India; so this is the reason why wild dog 'kills' in the former country are almost always found near water.

Seizure is made in several ways. Tearing bites at the flanks by which the animal is disembowelled; and, as Dunbar Brander says, seizure by the ears, nose, eyes, lips, hanging on like leeches, bearing down the head, quietly waiting and never letting go until the end. 'Robin Hood's 'lovely little russet-red balls of fur' well demonstrated this tenacious grip. No wonder few animals escape them. He, and others also, relate how a wild dog will cling to the back of a galloping animal and not be shaken off even when the heavy beast comes crashing to the ground.

The terrified heart-rending screams of victims of the hunt once heard can never be forgotten. It is Nature at its worst. The red fiends do not even wait until the prey is dead.

No doubt the testicles are sometimes seized. Dunbar Brander shows that this is accidental and not a habit. It is possible however that some dog may learn by accident the efficacy of such a hold. In Vol. XXXIII, at page 704, is an interesting account of pig-hunting in Java by 'native fox-red pariah dogs, '*gladak*', said to be descended from wild dogs possibly crossed with jackal, and 24 inches at the shoulder. These had evolved perfect team work, and one of the pack, always the same dog, invariably seized by the testicles which he removed in one rending mouthful.'

Mr. L.E.C. Hurst tells me that one of his pig-hunting hounds always seized a boar by this hold. At page 813 of Vol. XXXI Mr. Sálim A. Ali writes that he saw a sambhur just killed by a pack of wild dogs and strangely enough, the only part touched in the hind quarters were the testicles, which were clean missing. There are other cases also.

Seizure by the eye is common. The Chenchus of the Nullamalais affirm that wild dogs always seize game by the eyes if possible.

In some instances the eye is removed without damage to the lids. That is a fact and is due no doubt to eyes of deer being rather protuberant. Inverarity notes that the eyes are eaten immediately, but doubts whether the dogs seize at the spot. Since his day—1896—many observations have been recorded. At page 389 of Vol. XXXVIII Livesey notes that a sow had both eyes removed without damage to the lids or any other injury to the head. That supports Inverarity. At page 267 of Vol. XXVIII a Thamin was killed before the wild dogs had run into it; one eye was freshly torn out, the other badly gashed. That upsets Inverarity and supports the Chenchus.

Instances of single dogs successfully hunting deer are not infrequently recorded. They often hunt in pairs, but this is probably only during the breeding season. R.C. Morris notes that he once saw a solitary large dog watch a sounder of pig into a patch of tall sword-grass, and then follow the sounder into cover. It was soon chased out with loud noises by a wild boar; and on another occasion he saw a large solitary dog watching, and edging down to, a 'tat' pony that was grazing in the jungle.

Packs may number from a small number up to as many as forty, and perhaps even more. The Author of 'The Second Jungle Book' may have had good reason for writing 'The dhole do not begin to call themselves a pack till they are a hundred strong.'

Sometimes single dogs are met with, separate from the pack by some mishap, and these are no less care-free—'bold and saucy' as 'Hawkeye' aptly expresses it—in their demeanour than when in company. I have notes of three in particular. One stood by the road while the car was stopped to observe it at a distance of fifteen feet. Another, in U.P. was lame and came to a recent kill of a cow by a panther, driving off the vultures and tearing off chunks of meat which were thrown in the air and bolted whole. In the C.P. a lone dog came to a tiger's kill in the evening. He had a stumpy tail the end of which was completely healed. Bitten off in a fight? Damaged, fly-blown, and self-removed? I had seen him the day before and missed him with the 22 rifle. I saw no other dogs in that part of the Jarkahu Shooting Block. The first I knew of his arrival was a loud staccato hyaena-like noise under my tree. Very shortly the red dog crept out—and was soon at work. There was a marked difference in his now confident demeanour from that of a jackal in similar circumstances.

After breaking into the stomach from behind, where the tiger had made things easy for him, he deserted this unpleasant dish and attacked the flank where he soon had a six-inch piece of the hide off. It was wonderful to watch him make a small entry and then cut his way along with his back teeth (one short at back of each side of his lower jaw!) as neatly as one would do with a pair of scissors. I have watched a young tiger, also panthers, effect the same neat removal of skin. The dog left at dusk and returned at 11 p.m., by which time the tiger had been shot and was lying some way off. The dog

stayed a few minutes only and returned early in the morning. He chased away an intruding jackal—such a hasty pebble-scattering departure! I talked loudly to him without making him look up. He took no notice of my powerful whistle. Even when my signal horn was loudly blown to call up the men he merely jumped about in astonishment—not in alarm—and failed to locate the startling noise sounded from a short twenty feet above his head.

I have noticed the same thing with jackals, foxes, deer, and wild pig, all animals unaccustomed to look above ground level for their enemies. It has rather surprised me to read in JUNGLE DAYS, a book published in 1935, of a wild dog's behaviour at a tiger kill:- 'After ten minutes the dog displayed great cunning and knowledge for he began to look up systematically into each tree around him, and eventually caught sight of the machan and was shot.' And some time ago a writer of thrilling shikar yarns in an illustrated weekly, describing a buffalo-tiger fight, related how a jackal came to the kill from the neighbouring jungle, looked up from a distance of forty yards (I am not sure it was not sixty), saw the watching sportsman in his machan, and bolted into the forest. Such happenings are contrary to my observations.

Indifference to Pain

It has been observed by most of us who have shot wild dogs how indifferent to pain they appear to be. Never a sound will they utter however severe the wound, and will go miles with a body wound or a broken limb. Even when closely approached to put them out of pain they utter no sound and do not attempt to attack or bite. (page 389, Vol. XXXVIII).

Best writes in INDIAN SHIKAR NOTES that he has seen a wild dog turn and swallow his own protruding entrails, which had to be dragged from its throat with some force after the beast was finally killed. Phythian-Adams has seen a wild dog tearing at the guts of a wounded companion. I have read somewhere that a panther was seen to remove its entrails protruding from a stomach wound and impeding its progress. Many animals appear not to suffer pain from severe wounds. From my own experience I can say that sudden severe injuries do not cause pain; no doubt it is the shock to the adjacent nerves. But that is not always the case. A man shot through the stomach from behind at close quarters, the Martini Henry bullet missing the spine, suffered shrieking agony and tore at the extruded entrails with his hands. A tiger or panther with a stomach wound is fighting mad, though the latter is pretty mad on very slight provocation!

Fluctuating Population

Various writers and observers have remarked upon the apparently causeless fluctuating wild dog population in a given tract.

At page 162 of Vol. XLI Mr. Theodore Hubback writes:-

'It is possible that wild dogs have a period during which they progressively increase and then due to some unknown reason become scarce, again increasing until some unknown peak is reached. This phenomenon is well recognized among some species, the Ptarmigan of Alaska being a striking example. It is Nature's way of adjusting; and something of this sort may operate to keep wild dogs within limits.'

The wild dog must have a very wide range within any given tract of country, and no doubt move to more distant places when game has been too much hunted by them; but the fluctuations seem to be greater than can be so accounted for. It may be that distemper carries off a large number from time to time, but there must be also other causes at work. Perhaps it is that when food is plentiful the number in litters increases, and *vice versa*.

In a book *Work and Sport in the I.C.S.*, it is stated at page 144 that a District Superintendent of Police—Mr. Sandell—serving at the time in the Eastern Ghats in the Madras Presidency, died from hydrophobia the result of being bitten by a wild dog. The circumstances are not related. It is conceivable that rabies would be passed by jackals to wild dogs, but this is the only case of the kind to be found; so it is not likely to be of very common occurrence.

In 'The Origin of Species' I find:

'We may confidently assert,—that all animals and plants are tending to increase at a geometrical ratio,—that all would rapidly stock every station in which they could anyhow exist,—and that this geometrical tendency to increase must be checked by destruction at some period of life,' and 'In looking at Nature it is necessary never to forget that every single organic being may be said to be striving to the utmost, to increase in numbers; (with the exception, since Darwin's day, of civilized birth control peoples!) that each lives by a struggle at some period of its life: that heavy destruction inevitably falls either on the young or old, during each generation or at recurring intervals. Lighten any check, mitigate the destruction ever so little, and the number of the species will almost instantaneously increase to any amount. The causes which check the natural tendency of each species to increase are most obscure…. We know not exactly what the checks are even in a single instance.'

So, as Dunbar Brander observes, the causes of the serious fluctuations of wild dog population are processes of Nature outside our knowledge.

Does dog eat dog?

We know that in the snowy wastes of the Arctic and Antarctic regions sledge dogs will, under stress of starvation, kill and eat their companions. But can it be seriously put forward as an explanation of the fluctuations of wild dog population that they kill and eat one another, as the writer of a book on sport in Assam has suggested?

'Robin Hood's' wild dogs killed and ate pariah dogs. Mr. L.E.C. Hurst tells me an authentic instance of an officer going on service in 1914 who left his two Rampur hounds in care of servants who kept them chained up and not too well fed. Breaking loose one day they chased a pariah dog down the highway, killed and ate it!

Caton Jones (Vol. XVIII, page 195) saw a wild dog bitch return to a three-quarter grown dog which had just fallen dead from poison, and begin to drag it away. Probably it was the mother. He killed another wild dog in the evening, cut off its tail and a strip of skin from tail to ears, and left the remains on the foot-strip. Next day the dog had been taken away, apparently by wild dogs, as there were no marks of hyaena or panther.

Mr. D'Arcy McArthy tells me that two wild dogs he shot and left lying near the tiger kill on which they had been feeding, while he followed a wounded dog into the forest, had disappeared on his return and could nowhere be found. He concluded the dogs had come back, taken the dogs away and eaten them. 'I believe that they will finish off and devour any wounded members of their packs.' (Best, *Indian Shikar Notes*.)

It cannot be held from those few instances of cannibalism that wild dogs of India kill and devour one another as a habit when they unduly increase in number, or their normal food supply becomes scarce. There is no evidence of that: nor is it probable.

Conclusion

I have to ask Mr. Dunbar Brander to excuse my having so extensively made use of the Chapter on the Wild Dog in his book. There are some interesting comments and observations therein which have not been included by me.

Those interested in the Wild Dog should read this article in conjunction with 'Wild Animals in Central India'.

A fitting end to this article are the following lines from the pen of the master craftsman, author of THE SECOND JUNGLE BOOK:-

Red Dog

For our while and our excellent nights—for the nights of swift running,

Fair ranging, far-seeing, good hunting, sure cunning!

For the smells of the dawning, untainted, ere dew has departed

For the rush through the mist, and the quarry blind-started!

For the cry of our mates when the sambhur has wheeled and is standing at bay,

For the risk and the riot of night!

For the sleep at the lair-mouth by day,

It is met, and we go to the fight.

Bay! O Bay!

In the article where a Journal name is not given and only the Vol. No. of the Journal is mentioned the reference is to the Journal of the Bombay Natural History Society. P.Z.S. = Proceedings of the Zoological Society, London. – Editors.]

■ ■ ■

On Mimicry in Butterflies for Protection

By Col Charles Swinhoe M.A. (Oxon.) (Indian Army)

Year of publication: 1887

Col Charles Swinhoe, one of the eight founder members of the Bombay Natural History Society, was born in 1836 and at the age of 19 entered the Army as an ensign in the 56th Regiment of Foot. He joined the Bombay Staff Corps where he served until his retirement from the Army. Col Swinhoe was a man of many parts and obtained fame as an entomologist particularly with the completion of the magnificent work on Indian butterflies, *Lepidoptera indica*. He was the only Indian Army Officer recognised for his scientific work with an honorary M.A. by Oxford University. Col. Swinhoe's collection held 40,000 specimens of 7,000 different species.

He also served with the army of The Sind and in the Afgan War. The bird skins he collected were given to the BNHS.

That butterflies are to be found all over the world, clothed in colours and patterns closely resembling their surroundings, has been long known. Groups like the Satyrinae that are fond of shady places and live on hill sides and rocky dells are nearly always of a dull-brown colour; the Euploeinae that inhabit dark moist dells and live in the thick undergrowth of forests are all black; the Pierinae that fly about in the sun in almost any kind of climate are generally white or yellow; and the desert group of this family, the Teracoli, that mostly frequent barren sandy tracts in the hottest parts of the world have their white colours tinted and patched with most brilliant sun-spots of bright yellow and salmon colour; they only fly about in the hottest part of the day, and are very difficult to distinguish. Then there are the leaf butterflies, or Kallimas, and their allies, which, when on the wing, frequent the tops of high trees; their flight is very swift, and most of them are of large size. On the upper surface their wings are often brilliantly coloured, but underneath have the coloration and markings of various kinds of leaves, and when they settle, you see them vanish into a tree and become at once invisible. The common Indian form, *Kallima inachus*, for instance, a N.-W. Himalayan insect, generally settles amongst the dried leaves of a tree, and perching head downwards with closed wings so exactly resembles a dried leaf as to be invisible. Many of the Pierinae also mimic leaves on their under surface. The largest of them are the Hebomoias. I have only two species of this genus, *H. glaucippe*, from various parts of India—very plentiful in Bombay, on Malabar Hill—and the Nicobar species, *roeepstorffii*, and they both represent excellent imitations of leaves on their under surface. The subject, however, of the mimicry of one form of butterfly of another form was first brought clearly before the scientific world by Mr. Bates in an excellent paper which appeared in the 'Transactions of the Linnaean Society' for 1862, Vol. 33, p. 495, and subsequently Mr. Wallace brought many remarkable facts on this subject to light. It was observed by Mr. Bates that imitating species are comparatively rare, whilst the imitated are to be found in great numbers, the two sets living together. The imitated were for the most part brilliantly coloured insects,

Blue Tiger *Tirumala limniace*, Model

and he therefore concluded that they must be protected from the attacks of birds, &c., by some secretion or noxious odour, and this has now been abundantly proved. The principle of mimicry has been written about and argued out by many scientific men since Mr. Bates first brought the matter to light in 1862. I simply propose to show as many of the types of mimicry as I can from the examples out of my own private collection of butterflies. As to how one butterfly comes to mimic another for protection has been explained by many authors, and not always on the same theory; but I take it that Darwin's explanation that many species of Lepidoptera are liable to considerable and abrupt

Common Mime *Chilasa clytia* (form *dissimilis*), Mimic

Common Crow *Euploea core,* Model

variations of colour is the keynote of the whole mystery. Let us look at *Hypolimnas misippus.* The normal form of this butterfly is black, with large white spots on the wings; the female mimics *Danais chrysippus* in its coloration and markings, this butterfly being of a bronze-reddish colour. Now the male of *Hypolimnas misippus* is a very pugnacious insect and is very active, and has a remarkably quick flight, and is therefore capable of protecting itself; it is very good food for birds, lizards, &c., and whenever caught is a delicious mouthful; the female, however, is much slower in flight, and when heavily laden with eggs is easily captured. *Danais chrysippus,* on the contrary, like all the Danainae group, is a butterfly that no bird or lizard will touch, and both these species live in the same places. Now, supposing at some former period, in accordance with the well-known fact that *Hypolimnas misippus* in common with many species of Lepidoptera being liable to considerable and abrupt variation in colour (I myself have a very curiously coloured

Common Mime *Chilasa clytia* (form *clytia*), Mimic

Common Mormon *Papilio polytes* (female, form *stichius*), Mimic

female of this group), if a female appeared of a reddish or bronzy tinge (a not uncommon occurrence with black butterflies), would it not be probable that it would have a greater chance of escaping the attacks of birds and lizards than its black sisters? Some of its progeny would also probably have a bronzy tinge, and these also would have the greater chance to escape, and so on, from generation to generation the more bronzy the offspring became, and the more they resembled the colouration of the protecting species, the more they would become protected themselves, until, in the course of ages, the black form of the female *H. misippus* would cease to exist and its place would be taken by the beautiful female mimic of *Danais chrysippus*; and it is curious to observe that the protected and protecting forms are invariably found together. *Danais chrysippus* is an insect common in many parts of the world, all over India, Burma and Ceylon, in the Philippine Islands, in Turkey, Madagascar, Arabia, and the west, south, and south-eastern coast of Africa, and in all these places (I am not sure about Turkey) the protected form, *Hypolimnus misippus*, is also to be found. In Aden and in several parts Africa there is a form of *Danais chrysippus*, called *D. alcippus*, with white hind wings, and in all such places the protected form of

Common Rose *Atrophaneura aristolochiae*, Model

Plain Tiger *Danaus chrysippus,* Model

H. misippas is found with white wings; and in Aden, on the Kutch Coast in Sind, and in parts of the interior of Africa, there is a form of *D. chrysippus* called *D. dorippus*, without the black apical patch to the four wings, and in these places the female of *H. misippus* is also coloured and marked similarly. This form of the female of *H. misippus* is frequently to be seen in Bombay and other parts of India, and it is not at all uncommon, though not nearly so plentiful, as the *D. chrysippus* form. On observing this I have for some years collected all the *D. chrysippus* I could get together in the expectation of getting some *D. dorippus*, and in this I have not been disappointed, and I have now specimens in my

Danaid Eggfly *Hypolimnas misippus* (female), Mimic

ISAAC KEHIMKAR

Crimson Rose *Atrophaneura hector*, Model

collection from Bombay, Poona, Khandalla, and from the Punjab. It is, however, nothing like so common as the female of *H. misippus*, which mimics this form, reversing the rule that the imitating species are comparatively rare whilst the imitated swarm in large numbers; but this only shows that in former ages in these places the form *D. dorippus* was a common form, and that it has gradually been dying out and is now very nearly extinct. On the principle that mimicry is merely for protection, and that the protecting butterflies are those most abundant, we would here in India naturally expect to find the several species of the sub-families Euploeinae and Danainae more frequently mimicked than any other kind, because many of the species of both these sub-families are to be found in great abundance in most parts of India, and all are distasteful to birds, lizards, &c., and this is actually the case. It is very difficult to demonstrate facts of this nature for a private

ISAAC KEHIMKAR

Common Mormon *Papilio polytes*, (female, form *romulus*) Mimic collection from want of sufficient specimens, but happily my collection affords some very interesting examples, and though I cannot in all cases show the exact species mimicked, some of the mimicking species being from parts of India from which I have not many specimens, still I can show forms sufficiently allied to make the matter understood. We will first take the Euploeinae, of which the common form is *E. core*. It has many allies all over India, and its allies are more or less closely mimicked by several species of Papilio— *Papilio panope*, *Papilio clytia*, *Papilio lankeswara*, *Papilio dravidarum*, and the female of *P. castor*, also *Papilio tavoyana*, which exactly mimics *Euploea alcathoe* from the same parts of India, and of which I happen to have two good examples. There is another butterfly the female of which also mimics the Euploeas—a butterfly called *Hypolimnas bolina*, of the family Nymphalinae, widely separated from the family Papilioninae. Those are also some very interesting mimics of two other common species of *Euploea*—*E. midamas* and *E. rhadamanthus*. Two moths called *Amesia aliris*, which mimic the male, and three other moths called *Amesia midama* (all of the family Chalcosidae) which mimic both sexes of *E. midamas*; and also five excellent mimics, all butterflies of the family Elyminiinae, *E. leucocyma* and *Dyctis patna*, the sexes of which mimic the same sexes of *E. midamas*. *E. rhadamanthus* of both sexes and *Euripus halitherses*, a butterfly of the family Nymphalinae, the males of which mimic a *Danais*. The female mimics two forms of *Euploea*, *E. rhadamanthus* and a black *Euploea*. Next we will take the red Danainae *D. chrysippus*, *D. dorippus*, and *D. alcippus*. The female of *Hypolimnas* mimics all these, as before explained, and *Danais genutia* is mimicked by the females of three different species of the family Elymniinae, i.e., *E. fraterna* from Ceylon, *E. caudata* from South India, and *E. undularis* from Sikkim and Assam. There is a female of the last named species from Rangoon along with the allied form of *D. genutia* from that part of India, with white hind wings, called *D. hegisippus*, and it is very curious to observe that the hind wings of this and *E. undulais* are also whitish. In this case I also have another species of this family called *Dyctis vasudeva*, mimics a *Delias* of the family Pierinae, a gaudily-coloured common genus which nothing will eat. There are some white *Danias* mimicked by various kinds of *Papilios*, by one species of the famly Nymphalinae, *Hestina nama*, and by one species of the family Satyrinae, *Orinoma damaris*, *Euploea tytia* and *E. malaneus* beautifully mimicked by *Papilio agestor* and *P. govindra*; also *P. epycides*, *P. megareus*, *P. macareus*, *P. xenocles*, and *Hestina nama* of the family Nymphalinae, all of which mimic various forms of white *Danais*. Finally, there are some insects that mimic the common *Papilio diphilus* and its allies, a butterfly most distasteful to birds, &c., *P. pammon*, the female of which mimics two species, *P. diplhilus* and *P. hector*, and in the Nicobars the female of the variety *nicobarus* mimics the Nicobar variety of *P. diphilus*, called *P. camorta*. *P. janaka* is mimicked by a moth called *Epicopeia polidora*, of the family Chalcosidae, and *P. aidoneus* is mimicked by another moth of the same genus called *Epicopeia polinora*. If we examine the moths we find numerous cases of mimicry, commencing with the Zygaenidae, which mimic various kinds of hornets, wasps, and flies. There is another form of so-called mimicry, which is not mimicry at all. In the family Euploeinae there are many series of species which in their markings much resemble each other, but as they are all distasteful to birds, lizards, &c., there can be, in so far as we know, no reason why they should mimic each other; but, as has been already shown many of them are very closely mimicked by various other kinds of butterflies, some of which belong to families widely separated from each other and by many moths. All the Indian species of Euploeinae, except one, *E. andamanensis*, are coloured black, and it is undoubtedly a fact that many of them, though differing so much in the shape of their wings and in their sexual marks as to have caused their separation into different sub-genera, are so nearly like each other in

their markings as to be hardly distinguishable except to the experienced lepidopterist. These similarly marked species, in so far as I can understand it, must have had the same common ancestor, and for some reason unknown to us, though their markings have remained similar, the shape of their wings and the sexual brands on their wings have become altered in the course of time, to adapt them to their conditions of life in the great struggle for existence. It is also very curious to note how evenly these changes seem to have occurred in widely separated places, such, for instance, as in Bombay and Ceylon, where we have the common form, *E. core*, a black insect with largish white sub-marginal and marginal spots; it has the hinder margin of the fore wings nearly straight, and one small sexual brand on the fore wings of the male. We also get in Bombay *E. kollari*, so like it in its markings as to make it seem at the first glance to be the same insect, but if you examine it carefully you will see that it is quite different in the shape and size of the wings in both sexes, and the hinder margin of the fore wing is deeply curved outwards, and the sexual brand of the fore wing is also quite different. Now in Ceylon we have a form of *E. core* called *E. ascla,* also quite common there, differing from *E. core* in having all the spots small; and we also get *E. sinhala*, differing from *E. kollari* in exactly the same way that *E. ascla* differs from *E. core*. The *core* form is very common, and the *kollari* form is rare, and I believe the latter was the original form; that it is gradually dying out and has been replaced and pushed out of existence by the other, which has now become the common form. I cannot do better than to end this paper with a quotation from Darwin on this subject; he says:- "As in each fully-stocked country natural selection necessarily acts by the selected form having some advantage in the struggle for life over the other forms, there will be a constant tendency in the improved descendants of any one species to supplant and exterminate in each stage of descent their predecessors and their original parent."

Common Mormon *Papilio polytes* (female, form *stichius*)

Indian Sheep Dogs

Lieut Gen William Osborn, I.S.C. (Indian Army)
Year of Publication: 1901

After reading Note No. 23 in the *Journal of the Bombay Natural History Society*, published on the 18th of May 1901, on Indian Sheep Dogs (**see Appendix below**) I can fully endorse all that "J.F.G." has therein written on their instinct, courage, and training, as I have had during my travels, and shooting rambles, very many opportunities of seeing, and watching, the working of these valuable assistants to the Indian shepherd. That these dogs can, and do, drive off wolves, I think there is no doubt. I have seen a pair of wolves watching a flock of sheep, during the temporary absence of the shepherd. The dogs being on guard, the wolves were evidently afraid to attack, though everything was in their favour, except the Sheep Dogs. So intent were these two wolves on the business before them, waiting for a chance, that I was able to shoot one of the pair, the female.

From the large hairy sheep dog of the Guddis, who come down with their sheep and goats from Chamba, Lahoul, and Spiti, into the North Punjab during the winter, down to the sheep dogs of Southern India, these animals are nearly all trained in the manner described by "J.F.G." Of their ferocity, and capability of attacking any animal whatsoever that approaches their flocks, I once had an interesting experience. I was black buck shooting on the plains between Bellary in the Ceded Districts and Hurryhur in the Mysore Country. I had wounded a fine buck, and was riding him down with the spear. The buck was practically mine, for the plain extended for miles; my nag had plenty of go left in him and the buck was getting done, when unluckily for him, he took a course which led him quite close to a sheep-fold. Directly he passed it, three large sheep dogs bounded over the thorn fence, attracted by the sound of the buck galloping over the stony ground. At this point of the chase I was only thirty or forty yards behind. The dogs laid into the buck in first rate style, and pulled him down in about a quarter of a mile. I jumped off my horse, intending to give the coup de grace, but so fierce and determined were the dogs that I thought it most prudent to stand out, and let the fight go on without me. The buck was dead and mangled by the time the shepherds came up, and they rescued the venison for me. Had I interfered, I think I should have fared badly, especially as I am sure the dogs had never seen a *Feringhee* before. When the rally was over I gralloched the buck and threw the whole of the viscera to the dogs, to reward them for their assistance, and for the interesting piece of sport they had shown me.

Lieut Gen William Osborn was born in 1828 and joined the Army in 1848. He was an officer in the Royal Engineers, and was assigned to the PWD (military works) at Bombay when he was a Lieut Col. He was with the Indian Staff Corps from 1866 till the time of his retirement.

Tibetan Mastiff

These sheep dogs of the Deccan, the Ceded districts and the adjacent province of Mysore, are all of the same class, chiefly red in colour, a few black and tan, and a very few quite black. Many of the red ones are feathered on the ears, tail and down the forelegs, and there are many quite smooth, like the ordinary red pariah.

It is not all about sheep dogs, however, that I am writing; I wish to say a word or two on behalf of the common dog of the country, the unjustly despised Pariah. I don't mean the Mongrel that one sees about Indian towns and cantonments but the true Indian Pariah Dog, mostly red in colour.

That we have neglected this animal as a faithful companion, good watch dog, and an excellent assistant in many field sports, there is no doubt, though it is not strange that we should have done so, as sportsmen are a conservative body, many of whom consider that there is nothing good in the sporting line out of England. But of the good qualities of the true Pariah, as I have to call him, I have seen many instances. Notably when passing the hot weather months on the Ramandroog Hills, not quite 40 miles from Bellary, I found there were sixteen men of a tribe called "Bender" in the village below my camp who used to hunt with their dogs which were of the same class as I have described, the true breed of country dog from which the sheep dogs are taken.

These sixteen men had a pack of eight dogs. Each man was armed with a spear, a small axe, and a knife. In addition to these, he carried a flint and steel, and tinder in his pouch. I am writing of a time years ago, when there was a fair head of game on this small range of hills, consisting of tigers, and leopards, many sambur, pigs, etc. These Benders used to turn out for a hunt regularly twice a week, their game being always sambur, and in those times it was not long before the pack of eight were in full chase of a stag or hind. I never saw these dogs lose a sambur once. When they found they stuck staunchly to their quarry, and the end was always the same, stag, or hind, at bay, either against a rock, or in a pool of water, the pack laying around, and the sambur slain at last

by the spears of the Benders exactly, from start to finish as is described by Sir Samuel Baker in his description of sambur hunting with hounds, in his book THE RIFLE AND HOUND IN CEYLON.

I am not writing a sporting article but I am endeavouring to show the good qualities of the Indian dog. Sometimes these same "Benders" used to hunt hares in the grassy plains below the hills. Assisted by their eight dogs (all red ones) and armed only with their throwing sticks, a curved hardwood stick with a knob at one end shaped something like a boomerang, I have seen them bring home fifteen to twenty hares, not one of which they could have secured without their dogs.

Once I was after a man-eating tigress; two Benders and one of their dogs were with me. I wounded the tigress which took refuge in a deep rocky glen, thickly covered in with a species of climbing, thorny mimosa. Entrance through this network of hooked thorns was impossible to a man, but the dog, a red pariah, was able to crawl in, found the tigress, and bayed her incessantly for half-an-hour. When the dog got too close, the tigress would execute a charge with the usual music, but could not get home, as her back was injured. However, the dog stuck to his work, and I was able to mark the spot where the tigress lay by the moving of the bushes, and meeting each charge with a couple of barrels, at hazard, a lucky shot at last finished the business, and I bagged the tigress which I certainly should have lost but for the dog.

These dogs are trained by native shikaris to other kinds of sport. Once when duck shooting in Mysore country, I was seated on a hillock watching a flight of ducks on a sheet of water, when I saw a performance that surprised me. In a hole dug in the ground about twenty yards from the brink of the water was seated a shikari, well concealed from

Chippiparai

the birds. He had with him his old gun and a red pariah dog. His object was to attract the birds to within shooting distance. To accomplish this, every now and then, at fairly regulated intervals, he threw a lump of a thick kind of *chupattie* they eat in these parts, down to the margin of the water. The red dog would then jump out of the hole, run to the *chupattie*, eat it, and return at once to his master. This was repeated till the attention of the ducks was attracted and as it was continued, the flock swam gently on in the direction of the dog in that curious manner in which many birds will follow, and mob their natural enemy. At length coming well within range, bang went the old musket, and the shikari emerged from his pit to gather in the slain.

The interesting point here, apart from the performance of the dog, is the well-known habit of wild birds following their natural foes. In this instance the ducks evidently mistook the red dog for their enemy the fox or jackal. In English decoys this habit has been taken advantage of. The decoy man trains a small red dog to show himself at different points to the ducks on the water. These invariably follow the dog slowly till he leads them into the mouth of the decoy net, and onwards, till the birds enter the fatal chamber from which there is no escape. Here we have an Indian shikari following a practice that has been for ages in use in England. Did we learn this trick from the East? The Indian fowlers could hardly have got it from us.

Appendix

Sheep dogs in England are wonderfully trained, and their intelligence in picking out, and driving sheep, is almost incredible; but clever as they are they cannot hold a candle to the Indian sheep dogs I have seen in the province of Berar. These dogs are not trained by man at all, but, their masters, who belong to the Shepherd, or Dunger, caste–for every trade or vocation has its caste in India–have discovered what the dog will instinctively do under certain circumstances, and arrange that those circumstances shall occur. The following story will show how I came to find out about these sheep dogs, and the value the shepherds place upon them, which is entirely owing to their training, as they belong to no particular breed.

About the year 1863, the G.I.P. Railway was being made through Berar and one of the contractor's staff, a platelayer or inspector, I forget which, lived in a tent about a mile and a half outside the town of Akola, where I was. A big masonry bridge had to be built across the Morna river, which passes through Akola, and to reach this bridge from his tent, the inspector, as we will call him, had every day to pass a thorn enclosure, in which was kept a large flock of sheep. Outside the door of the enclosure, when the sheep were there, were always several large and very fierce dogs, which invariably came out at the man, until at last, to give them a lesson, he took his gun with him, and the next time the dogs attacked him he shot two of them. This, though almost in self-defence, led to serious trouble, and the inspector for one, and I for another, did not understand why so much importance was attached to the death of the dogs, as to all appearance they were very ordinary mongrels, though large and long in the leg. The case came up before the Deputy Commissioner, when it was discovered that the statements of the inspector and of the shepherd who owned the dogs, were practically identical. That is, the complaint was the defence.

The inspector, a Scotchman, gave a large sum of money to the owner of the dogs, when their value was explained to him, and it was then that I first learned how an Indian shepherd protects his sheep from wolves and other wild animals when feeding in the jungle by day, and from thieves at night, with the assistance of these large fierce dogs, that in the performance of their duty, will go at a man as willingly as they would at a wolf or any other wild beast, though, strange to say, away from the flock they become cowards at once.

At first I thought the sum of money given by the inspector to the shepherd excessively large, but afterwards I understood how it was almost impossible that money could make good to the man the loss he had sustained. In talking to him, I got him to explain to me how these dogs were trained, and why they were so very brave and fierce. He said that the custom of the shepherds was, when they wanted to train a dog, to take a male pup from the litter as soon as it could see, while they were careful that the father and mother were of as large and heavy a breed as possible. This puppy would be given to some ewe who had lost her little one, and was suckled by her. At first she had to be held down for the purpose, but in about three weeks time she would become accustomed to being the dog's wet nurse, and would take to the little animal. As soon as the puppy could run about properly, it would be allowed to run alongside its foster-mother to the jungle, where she would go daily with the flock.

Care was taken that the ewe in question should remain, owing to her peculiar duties, an unusually long time in milk, so the puppy was well nurtured in its infancy, and invariably turned out a larger and heavier animal than its parents, this being also partly due to the fact that he was castrated when about a month and a half old. From always going out with the flock, life among the sheep became the dog's second nature, and when it no longer got milk from its foster mother, the shepherd would feed it well with milk from the other sheep, and bread broken into the milk. And when from any accidental cause a sheep died, its carcase would be divided among the dogs of the flock. Finally, the puppy being full grown, would be turned out of the flock at night time, and would sleep outside the door of the thorn enclosure with the other dogs, generally round a fire made of dried sheep's droppings.

A large flock of sheep generally requires six, seven, or eight dogs, all educated in the way I have described. Every day after the sun has well risen and dried the dew off the grass, the flock is taken into the jungle to graze. As they start, one dog goes to each of the four corners of the flock, and remains about sixty or eighty yards from it. The rest of the dogs follow with the shepherd. Should a wolf appear, the dog at the corner nearest to him gives a bark, on which the dogs with the shepherd, who is always behind the flock, rush towards the warning voice to assist in repelling the intruder, but the other three sentinel dogs remain still at their posts. What is going on elsewhere is no business of theirs. They have only to give the alarm where they are, and the wonderful thing is that the shepherd does not tell the dogs to go to the corners, but the weaker ones invariably by their own instinct take this post, while the stronger and heavier dogs remain with the shepherd, on the alert, to rush at once to any spot to which they are called. As I have already said, dogs so brought up, when with their flock, are utterly fearless of man or beast; but if they leave the flock, be it only to walk with their master down a village street, they are the veriest cowards possible.

The shepherd told me that the daily routine of the dogs with the flock was due to no teaching of his. He said that it was instinct, and was quite unable to tell me where or when the system had been introduced. He only knew that from time immemorial it had existed. When I asked him if his dogs would really have bitten the inspector, he answered, "Of course they would", and added, "that no wise man would ever go near a sheep enclosure". And when I pointed out the fact that the inspector had been going along a public footpath, he replied "that his enclosure had been there before the footpath, which had been made by the workmen going to and coming from the railway bridge". The shepherd did not do so badly after all with the compensation he got for the loss of his dogs, as the increase of population and of cultivation had driven off the wild animals to a great extent from about Akola, so that he could manage to get along without those that had been shot; but I found afterwards that what he had said was quite true, he could not have replaced them under two years.

This account will show how cleverly the instinct of the dog has been utilized for the protection of his flock by the shepherd in Berar, and for all I know, also in other parts of India, and to me it is all the more wonderful, as from the circumstances of the training this cannot be an inherited instinct.

[The custom of rearing dogs for the protection of the flocks, amongst the sheep themselves goes back many hundreds of years. Darwin says that when staying in Bana Oriental he was amused with what he "saw and heard of the shepherd dogs of the country…. Their method of education consists in separating the puppy when very young from his bitch and in accustoming it to its future companions. The ewe is held three or four times a day for the little thing to suck, and a nest of wool is made for it in the sheep pen. At no time is it allowed to associate with other dogs or with the children of the family. The puppy is also castrated in order that it can have no common feeling with the rest of its kind. From this education it has no wish to leave the flock, and just as another dog will defend its master–man, so will these the sheep. It is amusing to observe when approaching a flock how the dog immediately advances barking and the sheep all close in his rear, as if round the oldest ram.' –Ed.]

■ ■ ■

Natural History of Deesa (N. Gujarat)

The famine years

By Capt C G Nurse (Indian Army)

Year of publication: 1900

During the past three years, with the exception of some eight months spent in the Panjab during the Tirah expedition, my head-quarters have been at Deesa, and it may, perhaps interest others if I give a short account of the sport and natural history of this delightfully sunny spot. During the months of December, January and February only is the climate pleasant; in the hot weather, i.e., from March to June, and again in October and November, one is baked, and during the rains one is boiled. But, hot as it is, I personally much prefer the dry heat, which rises sometimes during the hot weather to 120°, to the moist and enervating climate of Bombay. The life has some compensations.

I will begin with the Mammalia. The only monkey we see in a wild state is the Common Langur, and it is not nearly so numerous here as it is near Ahmedabad, where it is a perfect nuisance. The Lion has of course long ago disappeared from this neighbourhood; the last was, I believe, killed in 1878 near the village of Bhoyen, about two miles from Deesa. The larger *Felidae* have been more than usually numerous this year; the famine has driven all the animals, both wild and domesticated, towards the streams, there being no grazing elsewhere, and Tigers and Panthers have naturally followed them. Of the smaller *Felidae*, the Jungle Cat (*Felis chaus*) is common. The Mungoose, is, of course, abundant. I fancy there are two kinds, but I never killed one for identification.

The Wolf is common, and I have seen them within three miles of Cantonments. Jackals of course swarm, and, thanks to a sporting Colonel of Native Cavalry, who imported several foxhounds and beagles, have shown us good sport. Two kinds of Foxes occur, one with a black tip to its tail (*Vulpes bengalensis*), and one with a white tip (*Vulpes leucopus*), and both have given good runs occasionally. The Otter is fairly common; I once saw a whole family at Malana tank, about 18 miles from Deesa; there were seven or eight altogether, and they jumped into the water one after the other, so I had a good view of them. The Indian Sloth Bear of course occurs in the wooded country towards Mount Abu. Hedgehogs are common, but I fancy they all belong to one species. I once caught a specimen, and wanted to idenfify it. He, however, refused to unroll, so I chloroformed him, and finding him to be the common *Erinaceus pictus*, gave him his freedom as soon as he got over the effects of the anaesthetic.

Capt C G Nurse was appointed to the Royal Irish Fusiliers on 26th January 1881. He fought in Sudan and was awarded the War Medal. Posted to the Bombay Staff Corps on 10th March 1885, he acquired proficiency in the Persian language and passed Interpreter examination in Russian. In 1886 he was posted to the 17th Bombay Infantry at Ahmedabad and in 1893 to the 13th Bombay Infantry at Deesa. In 1908 he was appointed Commandant of the 33rd Punjabis.

Red Fox *Vulpes vulpes*

Indian Fox *Vulpes bengalensis*

I only once tried to identify a bat specimen, which proved to be the common *Nycticejus kuhli*. Flying foxes are not usually numerous, but in October and November, 1898, there seemed to an extraordinary quantity about. I presume that some fruit must have attracted large numbers from elsewhere, as I have never seen so many anywhere as I saw then. The common Squirrel (*Funambulus palmarum*) is a perfect nuisance to anyone having a garden, as it does considerable damage by eating off young shoots. The article in the 'Fauna of India' Series on this species seems to leave it doubtful whether it destroys birds' eggs or not. I have not the least doubt that it does so whenever it finds them unprotected, and on one occasion I purposely left an egg on the ground where it could be seen by a squirrel, with the result that it was sucked dry before my eyes in a few minutes. Rats and Mice are of course common enough, but I never tried to identify any of them. *A. gerbillus*, but whether *G. indicus* or *G. hurriance* I am not sure, is extremely common. I often wonder how they fare in the present famine year, but they seem as numerous as ever.

Flying fox *Pteropus giganteus*

The Porcupine is apparently common, but owing to its nocturnal habits, seldom seen. Hares are very numerous in some places; they all seem to belong to one species, the common *Lepus ruficaudatus*. Nilgai are extremely common and very tame, but are not allowed to be shot. During the past cold weather they have always seemed in good condition, although nearly all the cattle in the country have died of starvation. I have several times seen them feeding at night in jowari fields, notwithstanding the fact that the latter are fenced all round and generally watched day and night. This probably is the reason that they appear so fat and well, as there is little or nothing for them to eat in the jungle.

Blackbuck are scarce in the immediate vicinity of Deesa, but Chinkara are common enough, though somewhat wild. Sambar are numerous in the hills round Abu; I am told that many of them have died of famine during the past year. Chital I have not personally come across, but I believe that they are not uncommon in suitable localities near the foot of the hills. The "Mighty Boar" is not nearly so numerous as many of us could wish, and pigsticking, which once flourished in this neighbourhood, is at present at a somewhat low ebb.

As regards Birds, Butler's list in 'Stray Feathers' of the avifauna of this neighbourhood is pretty exhaustive, and I made no attempt to add to it. Birds, in fact, appealed to me more as a sportsman than as a naturalist. Though I had heard of Bustard on several previous occasions, I did not come across them until the past cold weather, when I saw two. Houbara were also numerous this season, though they had hitherto been very scarce.

I came across a shikari who was a perfect artist in driving both the species in whatever direction he wished. Leading a camel, and walking in a circle, he would leave the guns behind any convenient bush, and then proceed till he got to the far side of the game. He would then begin to walk in a zigzag direction towards the guns, driving the birds, which would almost invariably fly overhead well within shot. Florican are fairly numerous at the beginning of the rains. I always have some qualms of conscience about shooting both Florican and Rain Quail at this season, but one is glad of a change from inferior mutton and bazaar murghi, and I fear that it is with me as with many others a case of —

"Video meliora proboque, deteriora sequor."

Of Sandgrouse we usually get only two kind in the immediate vicinity of Deesa, viz., the Common and the Painted. But during the past cold weather, the Large Sandgrouse appeared in considerable numbers. Peafowl of course swarm, but are considered sacred. Jungle and Spur Fowl are pretty common at the foot of the hills. The Grey and Painted Partridge are both fairly abundant, and six kinds of Quail may be obtained, though, from the sportsman's point of view, only the Grey and Rain Quail are worth taking into consideration. The former appear in countless thousands in September; large numbers remain till the end of October, when most of them apparently go south; they reappear about the end of December. The Rain Quail appear from about the middle of June, and are plentiful enough till the end of July. After this time they are not so much in evidence, as the grass has become fairly high by the beginning of August, and they are thus able to breed unmolested, so far as their human enemies are concerned.

Houbara Bustard *Chlamydotis undulata*

A few Rails, which I have not taken the trouble to identify, complete the list of game birds.

The cold weather of 1899-1900 has, owing to the famine, been an abnormal one on this side of India. Most of the usual migratory birds have scarcely appeared at all, or have come in greatly diminished numbers, and birds of prey have consequently been much fewer than usual. I have not seen a dozen Grey Quail during the whole of the cold weather. Duck and Snipe, however, which in ordinary seasons are few and far between in the immediate neighbourhood of Deesa, have frequented every likely and unlikely spot in the river Banas, although the latter is in most places only a few yards wide and a few inches deep. Demoiselle Crane, too, appeared in large flocks in October, flying up and down the river seeking food, but after about a month they disappeared, and I have not seen them since.

The non-migratory birds must have had a very poor time, and it appears marvellous that thousands of them have not perished of hunger. The struggle for existence during such a year as the present one must be terrible, and I often wonder how some species find any food at all. Except in Cantonments, and along the bed of the river, where there is a little cultivation, there is scarcely a green leaf or a blade of grass or corn for miles, and yet every morning shortly after sunrise, and every evening about sunset, enormous flocks of the common Rose-ringed Paroquet may be seen, leaving or returning to the trees where they pass the night. Where do they all obtain food? There are no wild fruits, and the little grain that is being cultivated with the help of irrigation is carefully watched and guarded.

Rose-ringed Parakeet *Psittacula krameri*

Among the Reptiles, I need hardly say that the common "Mugger" (*Crocodylus palustris*) is fairly abundant. Of the *Chelonia* I have only come across *Testudo elegans*, which is very numerous during the rains in the grass bhirs. The commonest house Gecko, appears to be *Hemidactylus leschenaulti*. On one occasion I saw one make a dash at a feather, which was blowing along the floor, mistaking it for an insect. Another time I was setting insects, and I accidentally dropped one on to the floor; before I could pick it up it was swallowed by a gecko. I then purposely dropped a small bee, with a short entomological pin through its thorax. This was also swallowed and the Gecko seemed much astonished at the pin, and made several unsuccessful attempts to get rid of the insect. He probably had severe indigestion for some time afterwards.

Among the Lizards, a species of *Varanus* is common. The lowest caste native eat its flesh, and make drumheads of its skin. The so-called "Blood-sucker", (*Calotes versicolor*), is extremely common, but I have scarcely ever seen one in the cold weather. I presume they hibernate in holes in the ground. The commonest Lizards are *Sitana pondiceriana* and *Agama minor*; the latter may generally be seen sitting outside its holes in the evenings during the hot weather.

Snakes are fairly numerous; the Cobra swarms in the grass bhirs in the rains. The Russell's Viper is common, and also *Echis carinata*. One of the most abundant snakes here is *Boiga trigonata*, which, as Boulenger says, bears an extraordinarily superficial resemblance to *Echis carinata*. I have frequently seen *Boiga trigonata* curled up on the top of cactus hedges. The elegant *Psammophis leithii* is also common, as are two, if not three, species of *Zamenis*. The plebeian-looking *Eryx johnii* occurs, and I obtained a specimen of some species of *Typhlops* from beneath some rubbish in my garden.

Starred Tortoise *Geochelone elegans*

Frogs and Toads I have not attempted to identify, and the same may be said as regards the few species of Fish that are obtainable in the neighbourhood. We get Murrel and Mahseer occasionally, but they generally have a muddy taste, and the only local Fish which is, in my opinion, worth eating is a so-called "country whitebait", but I have not the least idea to what species this belongs.

About 50 species of Butterflies occur, chiefly in the rains. The most interesting are, perhaps, the various species of *Teracolus*, which is chiefly, if not exclusively, a desert genus. Some half a dozen species occur here, and some of them positively swarm during the rains. Moths are fairly numerous, but I have not yet attempted to identify those I collected, though I obtained a fair number.

Common Cat Snake *Boiga trigonata*

One amusing incident I recall; one evening at the beginning of the rains several death's-head moths flew into the room, and settled on the ceiling. There were a great many geckos about, most of them with their abdomens considerably distended from the number of small insects they had consumed. A gecko, bolder than the rest, rushed up to one of the death's-head moths and seized it by a leg; another rushed up from the opposite side, and seized another leg. Then commenced a tug of war, which ended in the moth flying away, and both geckos falling on the floor. I could not see whether the moth got off without the loss of a leg or not.

When I first arrived at Deesa, I noticed that there seemed to be more *Hymenoptera* than any other order of insects, and though I had hitherto paid little or no attention to this branch of Entomology, I determined to collect and identify as many species as possible. Bingham's volume dealing with a portion of the *Hymenoptera* in the FAUNA OF INDIA Series had just been published, and the author kindly assisted me when I was in doubt, and described in this *Journal* some new species obtained by me. Since then I have found the study of this order of absorbing interest, and have devoted a considerable part of my spare time to the collection and identification of specimens. Even in this barren locality I have succeeded in obtaining well over 150 species, not including Ants or *Hymenoptera parasitica*. A large proportion of these are apparently new species, and have yet to be described.

The parasitic Hymenoptera are not numerous, except the *Evanidae*. This genus is supposed to be parasitic on *Blattidae*, (Cockroaches, & c.), but I once bred a species of *Evania* from a larva of *Teracolus pleione* at Aden, so some of them are evidently parasitic on Lepidoptera.

I collected a fair number of Diptera, which I sent to England to be identified. One of the most interesting was the Horse Bot-fly, which I bred from larvae passed by a horse. I do not yet know if it is the same species that occurs in Europe.

Among the Orthoptera two species of Locusts are fairly common: a reddish and a yellowish kind. The latter sometimes arrives in small swarms at the beginning of the rains; but fortunately we have not had any large swarms in addition to our famine troubles. The so-called "milk-bush" (*Calotropis gigantica*), which is extremely abundant here, is frequently stripped quite bare by a species of locust, but this does not appear to be migratory, and so far as I am aware, does little or no damage to other plants.

White ants are only too numerous. Dragonflies are plentiful, and it has always been a puzzle to me where they can all come from in such a dry locality. They breed, of course, in water, and, though there is only one small stream here, and a few wells, yet at whatever time of the year I go into my garden, I can always see several species. Ant-lions are plentiful, and their pitfalls may be seen almost anywhere. I notice that the Cambridge Natural History states that "The *imago* is considered to be carnivorous." This I can confirm, as I have frequently seen a species of *Myrmeleon* common at Deesa catching small moths and beetles round a lamp at night.

Spiders do not seem to be very abundant, especially the larger kinds. The little red velvety species which appears at the beginning of the rains is one of the most striking. I have been told that a decoction of these is used by natives in Kathiawar, and possibly elsewhere, as an aphrodisiac.

In conclusion I have to say that I have been able, thanks to a taste for Natural History to pass many a "Long, long, Indian day" without boredom, even in the hot weather. I hope, later on, when I have had an opportunity of comparing my Hymenoptera with those in the British Museum and other collections, to supplement this somewhat discursive paper by a more scientific one, in which the new species collected will be described.

■ ■ ■

The Tiger – A Natural History

By Brig Gen R G Burton (Indian Army)

Year of publication: 1933

The Tiger, generally represented as savage and bloodthirsty, and in fact as the embodiment of those qualities, possesses characteristics common to the whole species although some individuals have a character and habits peculiar to themselves. He is no more bloodthirsty than carnivorous man, except that he does his own killing while man usually has his killing done for him, amounting in Great Britain to 40,000 creatures slaughtered daily for food. The animal killed by the jaws of a wild beast probably suffers no more than one done to death with a pole axe or even with a "humane killer"; the yells of a pig killed within my hearing sixty years ago have never been forgotten.

The Tiger is ravenous and ferocious in the attack on his prey, which he carries out with ruthless violence, and he is often savage when disturbed at his kill, as the dog may be when engaged with his dinner, although having the advantage of age-long contact with man. The Tigress will fiercely defend her young, but no one would blame her for the most savage exhibition of maternal love or instinct.

In his attitude towards man the Tiger in general displays a desire to avoid contact or close acquaintance, and even the man-eater shows timidity in hunting human beings. Such apparent timidity is not, however, always due to the cause of fear usually assigned to it. It is the wild beast's stealthy and unobserved approach to any prey for purposes of surprise. When wounded he usually exhibits the utmost courage and ferocity in retaliation on his enemy in defending himself from death or further injury. So does the human being; and it always seems to be a mistake to apply the term "cowardly", as is often done, to the animal that attempts to escape; although, even when wounded, the wild beast, like man, may sometimes endeavour to get away rather than face an enemy.

Undoubtedly character may vary with the individual. Particular Tigers acquire a reputation for ferocity or timidity with the inhabitants of the neighbourhood over which they range, and to whom in course of time they may become well known. I killed an old pair within a mile of a village in the depths of the jungle where the people spoke of them with respect as familiar objects, and neither feared them nor objected to their presence. "Sahib!" said the headman of the hamlet, "we have known these Tigers for more than a

Brig Gen R G Burton, the fourth son of Gen. E.F. Burton of the Madras Staff Corps, was commissioned in 1884 in the 1st West Indian Regiment and served in Jamaica. He was later appointed to the Bengal Staff Corps and served in various formations. He was Commander of the Cadet College at Wellington in 1917 and Commander of the Madras Brigade Area in 1918. He was a man of literary tastes and wrote books on military history and natural history. He died in 1951 in his 87th year.

VIVEK SINHA

dozen years, and they have never harmed us. Certainly they have killed some of our cattle, and we have seen them close to the village, but they have not attacked or molested any of us." Unarmed villagers would even drive the beasts from their prey, and secure the hide and flesh for themselves.

Often the Tiger is so timid that a little herd-boy may drive it from his flocks, but there is always danger of retaliation. In one instance a young Maratha woman named Parvati was tending cattle, when suddenly a Tiger came up to attack them, having on previous occasions been successful in carrying one of the animals off. The woman stood between the cattle and the beast, which then seized her by the left shoulder, breaking in pieces the bones of the upper arm. Having a bamboo staff in her right hand, the brave woman struck him over the head several times, and he let go and retreated into the jungle, leaving the cattle unharmed. She was taken to hospital, where the arm was removed at the socket. She recovered in two months, when she was given a reward of fifty rupees by the State of Sawantwadi, in which the incident occurred.[1]

---

[1] Related in the *Journal of the B.N.H. Society.*

But many Tigers are dangerous to approach when on their kill, and will roar and rush at the intruder, though probably not charging home with intent to kill, but merely to terrify and prevent interference. Sometimes a Tiger, although not a man-eater, if it has suffered an old injury or wound, will become ill-tempered and may hold up travellers and bullock-carts, apparently connecting human beings with the injury; while a recently wounded one will kill anyone who approaches it. Thus a Bombay sportsman wounded a Tiger which then killed a herd-boy; he found the dead boy, and on the spot was suddenly attacked and himself killed by the animal.

It used to be said that the Tigers of the Pench River in Nagpur were fiercer than in other parts of the country, but this may certainly be doubted, and no grounds have been stated for the supposition. No doubt the natives would ascribe an unusual number of accidents to a peculiarly ferocious habit of the animals of the district. This is quite apart from the. prevalence of man-eaters in particular districts, such as the Sundarbans of the Gangetic delta of Bengal, and formerly the neighbourhood of Bombay. The temper of many Tigers may have been tried by frequent interference or molestation, such as being hunted by sportsmen and driven from one cover to another. Here, and in the case of animals wounded, arises the question of the powers of memory of animals, and of their ability to think or to connect cause and effect.

We know that the memory of some domesticated animals such as dogs and horses is very remarkable. Whether wild beasts possess similar faculties in a greater or lesser degree

is more doubtful. But all who have hunted Tigers are aware that an animal that has often been beaten out or driven from cover is difficult to bring up to the guns posted in the usual manner. Yet, unless it has been wounded, it can scarcely be supposed that the animal reasons that it is being pursued for its life. Probably it merely fears the unknown and unusual. A Tiger will kill a beast out of a herd near which it is passing when being pursued, just as a hunted fox will snatch a fowl from the farmyard. Yet we constantly read instances described as "boldness" of Tigers or leopards returning to their prey soon after being fired at and missed, perhaps two or three times during a few hours, for human beings are prone to ascribe human ideas to animals. It is not probable that an animal, unless hit, connects the report of the rifle with an attempt on its life, or an attempted injury of any kind; nor can it be aware that a projectile has been propelled to its address, even though it may hear the whizz of the bullet. It has merely been subjected to an unwonted noise, which alarms it and is connected with human agency if a human being is in sight.

This is no doubt instinctive dread of the unknown and unfamiliar, but it is difficult to say where instinct ends and reason begins. Professor G.J. Romanes defined as instinctive—"actions which, owing to their frequent repetition, become so habitual in the course of generations that all individuals of the same species automatically perform the same actions under the stimulus supplied by the same appropriate circumstances". But rational actions are those required to meet unusual circumstances, and hence require intentional effort of adaptation.

RAVI SINGH

Instinct, we are told, is reflex or non-mental action into which there is imported the element of consciousness. While reason or intelligence is the faculty concerned in the intentional adaptation of means to ends. In fact, it implies some consciousness of cause and effect, themselves the two sides of one fact.

The habits of the Tiger in tropical and sub-tropical regions are nocturnal, apparently partly but not wholly owing to climatic reasons, but also to the value of darkness for concealment while hunting their prey. But in India in the cold weather, though rarely in the hot season, the Tiger may seize its prey by day, and may sometimes, though rarely, be seen on the move at all hours. It would be interesting to know whether this habit is confined to tropical and sub-tropical regions and whether the Tiger of Northern Asia is active by day as well as by night. Eyes would not be adapted to seeing in the dark except for such purposes of vision. .

While the Tiger displays great impatience of heat, retiring by day to shady cover, generally near water, and panting heavily with lolling and dripping tongue when driven from cover, it follows that it is also impatient of thirst. It has often been said that Tigers and peacocks live together in the same thickets, but this is because they both suffer from thirst, while the neighbourhood of water is naturally cooler and harbours cool and shady retreats. They sometimes inhabit caves, not only the recesses of rocky and basaltic heights such as those forming a conspicuous feature of the country between the Krishna and Godavery Rivers, but genuine caverns in the sides of hills, which form suitable retreats for the accouchement or for the habitation of Tigresses with their young cubs.

Sunderbans Tiger

Tigers are addicted to lying in water, and more than once I have seen them driven forth dripping with water after emerging from the bath in which they have been lying immersed. Some are more than others fond of this habit, but generally it may be said that where most of the cat tribe are averse from entering water, all Tigers have no hesitation in plunging into stream or sea, and all swim well.

Tigers were at one time very numerous on the scene of the battle of Plassey, Kasimbazar Island, where, Williamson wrote in his *Oriental Field Sports* early in the last century: "Tigers not only resort freely to the water when pursued, swimming in a manner that denotes their familiarity with that element., but may frequently be seen crossing large rivers, when no object appears to be in view. About Daudpore, Plassey, Agahdeep and especially along the borders of Jellinghee, they are known to cross and recross during the day as well as by night; seeming to consider the stream as no impediment. From Agahdeep, in particular, they pass over to the extensive jungle of Patally, which has ever been famous for the number it contained. I have, in passing through it, seen four several Tigers within the space of two hours", and a gentleman who was travelling in his palankeen in the year 1782 saw three lying in different parts of the road as he went on.

It is only to be expected that in the Sundarbans they should take to water, for the swampy delta is split up into numerous islands, and is everywhere traversed by many water channels. That region has always been a notorious abode of man-eaters, which become exceedingly bold and have been known to swim at night to a boat anchored in the stream, and carry away a man.

Mr. W.A. Hickie, who shot during several years in the Sundarbans, contributed interesting particulars regarding wild animals and water to the *Journal of the Bombay Natural History Society*. He found that Tigers take readily to water, and in some instances swim considerable distances up to three or four miles, even in tidal rivers with a four or five knot tide running during spring. He was struck by the intelligence shown by them in choosing their time for swimming, which was invariably at or about high water, when they are able to take off and land on hard ground. At all other states of the tides one has to flounder up several yards of bank through deep mud.

A curious feature of the forests is the entire absence of fresh water except in cultivated tracts, there being nothing but salt water to drink daily during the dry season from November to May. Fresh water is obtainable in the islands in certain suitable localities, but one has to dig for it four or five feet deep, the only way of obtaining it during a prolonged stay in or around the uninhabited islands on the sea front. These fresh water holes are soon discovered by the animals, which go mad after the water and flock to it to quench their thirst.

A number of instances is recorded of Tigers climbing trees. In some cases they have pulled down a man from a position of supposed safety, for it is generally considered that

VIVEK SINHA

there is safety at a height of twelve or fourteen feet from the ground. But a wounded one took a Bhil boy out of a tree from a height of twenty feet, springing up and making use of its claws. It grabbed the boy by the ankle, and the combined weight of the two broke the branch on which the boy was sitting and brought them to the ground; the boy's leg had to be amputated, but he recovered. In another instance, during a beat for a Tigress, a native "stop" in a tree called out: "here she comes," thinking himself quite safe. But she heard him, went up the tree, pulled him down, and bit him so severely that he died in hospital soon afterwards.

Usually the Tiger in a tree has been a young or a wounded animal, or often a female, probably not weighing over 250 pounds; it is scarcely possible that an old or heavy beast weighing perhaps double as much as the Tigress, could accomplish such a feat. A sports man who was on elephant-back beating the jungle started three Tigers. Suddenly the mahout called out: "Look, look, sahib there are three Tigers going up a tree!" The sportsman kept looking at the foot of the tree and surrounding jungle, never thinking of looking higher up, and before he saw them they were down again. He observed the marks of their claws as high as twenty feet from the ground.

It is generally supposed that the Tiger has no dangerous enemy but man, and no friend but those of his own species, as one would expect of a monstrous and savage beast possessing an armature unequalled except by that of the lion. But the incident of the Ootacamund Tigress climbing a tree to escape the yapping dogs, and the behaviour of lions when hunted in Africa with dogs, as Gordon Cumming hunted them with comparative safety to himself, and the fact that leopards have been seen treed by wild dogs, tend to confirm the possibility of the truth of the native stories prevalent all over India of Tigers being attacked and even killed by these animals.

The oldest book on Indian sport, Williamson's *Oriental Field Sports,* has an imaginary picture of wild dogs attacking a Tiger. The dogs as depicted are also imaginary, not bearing any resemblance to those animals. There is no record of one having been seen by a European attacked by wild dogs, and native stories of wild beasts have always to be received with caution, but it is easy to believe that they may be driven from their prey by these animals, and possibly even killed; for they have often been brought to bay by packs of curs, especially in the south of India. I have myself seen a Tiger put to flight and chased by a bull-terrier, which eventually brought the beast to bay but was mortally wounded in the encounter.

A native belief, prevalent throughout India, is that a pack of wild dogs will not hesitate to attack a Tiger, and that the latter generally gets the worst of the encounter. The villagers about the Katkamsandi Pass on the old Calcutta-Benares Road, which was formerly infested by wild beasts, many years ago related a story of having witnessed a fight between a Tiger and wild dogs, in which the dogs were victorious although they suffered heavy loss. Baldwin in his book on the large game of Bengal, relates a similar tale. In this instance the inhabitants of a small village in the wilds one moonlit night heard what they thought was two Tigers fighting. In the morning they came by chance upon the scattered bones of a Tiger recently killed, including one hind leg with a large piece of flesh adhering to it, and on the spot found three dead wild dogs; but there were not eaten because wild dog does not eat wild dog. A careful enquiry appeared to prove that the story was authentic. At Bitergaon in Berar, in May 1895, I was told that a Tiger had lately been attacked and its stomach torn out by wild dogs. It was said that a native official tried to have the beast driven out of the cover in which it was lying, but it charged and killed one of the beaters, and made off and was not seen again.

In an article in the *Pioneer* newspaper in 1895, a sportsman related that an aboriginal Chenchu woman, while picking mohwa blossoms, saw a pack of wild dogs in pursuit of a Tiger. The Chenchus, on receipt of the news, at once set out to bring in the Tiger, assured that it would be killed; they tracked the chase into dense jungle where it had escaped. When it was suggested that it had beaten the dogs off, the Chenchus said that in that case some dead dogs would be found. The winter afterwards the same sportsman was afforded what he calls "pretty practical proof" that the dogs do attack and kill Tigers. He wounded a Tiger, and a fortnight afterwards heard that it had been found dead fifteen miles off. He went to the spot and got the skin, skull, and claws from the people, who said that it had been killed by wild dogs, apparently in order to prove their claim to it; his reply was that if such were the case, the wound enabled them to kill it. It may be remarked that the wound he inflicted on the Tiger must have been slight, or it would probably not have travelled so great a distance.

VIVEK SINHA

Next week another was found dead, and a third some days later, both of which the Chenchus declared had been killed by wild dogs; and as no one else was shooting in the forest, and he himself had fired at only one Tiger, he was convinced that the story was true. The skulls and claws were brought to him, but the bodies were so decomposed that it was not possible to see if they bore the marks of bullets. The Chenchus apparently possessed no arms except perhaps bows and arrows, as I observed on a visit to the same hills. It is not likely, though not impossible, that they would attempt to kill Tigers with bows and arrows; in other parts of India they have been killed with these weapons, and one has been shot with an arrow head imbedded in its back; the skin over it had healed and the only outward evidence of the wound it had made was in a patch of lighter-coloured fur. Poisoned arrows might be used; Captain Forsyth related that two natives killed a man-eater with bow and arrow; Tigers have been killed with axes and spears, though it is doubtful whether the timid Chenchu would be equal to this.

A favourite illustration in some old books is that of a Tiger seized in the jaws of a crocodile, half immersed in a river or a lake, into which it is trying to drag its enemy; or of a struggle between the two. Whether such encounters are authentic or not, it is difficult to say, but it is conceivable that a Tiger might be driven by hunger to attack a crocodile, or to seize a young one as readily as it will a large lizard or a pangolin; and it is not impossible that one when drinking might be seized with relentless grip by the reptile; as cattle are attacked when they are drinking; the crocodile never lets go of its prey.

Among the enemies of the Tiger must be included not only the wild but especially the domesticated buffalo. A herd would probably not attack the Tiger unless provoked by its seizing a calf or other member, or, as has sometimes occurred, in protecting the herdsman. In such cases they may charge the enemy and gore or trample it to death. But a Tiger has been known to charge along the backs of a whole herd and kill the two men driving them. It will not often attack a full-grown buffalo, wild or domesticated, but instances have been recorded. In Cooch Behar a magnificent bull was seen moving leisurely along a river-bed on the border of Bhutan with a Tiger escorting it on either side. Every now and then one of them would rush in and try to get a hold, and the buffalo would merely sweep his horns. This went on for half a mile, when one of the Tigers got too close and the buffalo ripped it right up with his horns. It died at once, and the other bolted while the buffalo went on.

The Raja of Gauripur in Assam wounded a Tiger which got mixed up with a herd of tame buffaloes; when they had done tossing it about on their horns, there was little of it left. Nor is it likely that the full-grown gaur or Indian bison is often killed by a Tiger, though it is not perhaps as formidable an enemy as the buffalo. But I have found evidence of such a tragedy in the remains of a cow bison, where tracks showed that the herd to which it belonged had stampeded. A Gond shikari gave me a graphic description of a Tiger attacking a bull bison, which kept it off with its horns, while in another jungle I saw the head of a fine bull, killed after a prolonged struggle; the Tiger, when shot soon afterwards, was found to have an eye gouged out and other injuries. It is probable that the attack on one of the great oxen is made from behind, and the buffalo or bison disabled by ham-stringing.

The wild hog is a favourite prey of the Tiger, and a boar will no doubt sometimes put up a good fight, though a fine pair of tusks and a skull picked up from a kill prove that sometimes at any rate a very big boar is killed and eaten. Probably the boar would have little chance when taken by surprise and seized by the back of the neck. Encounters between the two animals have been described. In one instance in the Himalayas the snow was found to be trampled down over a considerable space, and covered with blood and hair, but the pig had effected his escape, having gone off one way and the Tiger the other, while drops of blood on the tracks of the Tiger showed that he had been marked by the boar's tusks. Walter Elliot found a full-grown Tiger newly killed, evidently by the rip of a boar's tusk, and in another instance villagers brought in a large one and a boar which had killed one another. The Tiger had jumped on the boar's back and seized it by the nape of the neck, but was killed by a gash in the stomach. Boars have been found to have the handwriting of the Tiger on their backs.

The Indian sloth bear does not escape the hunger of the Tiger, but can scarcely be numbered among his aggressive enemies, although a bear was seen to attack and begin biting a dead Tiger which had just been shot. Bears no doubt not infrequently fall victims, and I recollect the inhabitants of a

BALJIT SINGH

village in the Melghat forest bringing to camp the skin of a large bear which they said they had killed after it had been badly mauled by a Tiger with which it had had a prolonged struggle. The skin bore tooth and claw marks, and there was no reason to doubt the truth of their story; but the bear could probably not put up much of a fight against so formidable and agile an antagonist. I have found remains of bears eaten by Tigers, and have also beaten both animals out of the same cover.

A bear was seen by Sir S. Eardley-Wilmot to gallop through a jungle and stop on reaching a road. A Tiger appeared on the road fifty yards off, and began stalking the bear, stopping when the bear stopped, and sinking on its belly when the bear moved on. This went on until only ten feet separated the two animals. Meanwhile two other Tigers came out and lay down to watch. The bear left the road and entered a patch of grass, the Tiger followed, but turned away when the bear rushed roaring at him. In another beat a wounded Tiger rushed at a bear with a cub on her back; the bear made off yelling, and another bear, where a Tiger was in the beat, howling dismally, fled to the hills.

There is an animal commonly supposed to be a friend of the Tiger, and for whom indeed the great beast provides many a meal. This is the jackal, at one time supposed to be specially attached to the Tiger, and even thought by some to be not a jackal but an animal of some other species. Known in some parts of India, particularly in the west, as the "Kol-Bhalu", and elsewhere as the "Pheall" from the peculiar cry it utters in certain circumstances, the first mention of this animal is contained in a passage in Johnson's *Indian Field Sports* (1824). Johnson wrote that at midnight, when he was in a machan or tree-ambush, watching for an expected Tiger, he "heard at distance a Pheall, which is a jackal, following the scent of the Tiger and making a noise very different from their usual cry, which I imagine they do for the purpose of warning their species of danger… soon after the Tiger passed within a few yards of us; and, although we heard him distinctly purring as he went along, like a cat that is pleased, we could not see him in consequence of his keeping in the shade of the bushes. In a minute or two after he had passed, we plainly saw the jackal, and heard him cry when very near us… I have often heard it said that the Pheall, or 'provider' as it is commonly called, always goes before the Tiger; but in this instance he followed him, which I have also seen him do at other times. Evidently his cry is different from what it is at other times, which indicates danger being near; particularly as, whenever the cry is heard, the voice of no other jackal is heard, though at other times of the night they are calling in all directions: nor is that particular call ever heard in any part of the country where there are no large wild beasts of prey. Pheall, I believe, was the original and is now the proper name; but they are better known in Ramghur by the name of *Phinkar*, which in my opinion is more appropriate, as it explains what it is—'crier', 'proclaimer' or 'warner.' The former word was used from its resembling the cry they make, as in so many instances of the names of animals in this country"; and, he might have added, of animals in all countries and languages.

This is a very fair account of the pheall, but Johnson is mistaken in saying the cry is uttered only where there are large wild beasts of prey. Those who have spent much time in camps and observed the sights and sounds of the jungle must have heard not infrequently this peculiar call made by jackals not only near a Tiger or leopard, but evoked on other occasions, as when the jackal is pursued by dogs; while Sir S. Eardley-Wilmot observed two jackals gazing up into a tree, in which there was a large python, and uttering this cry.

Popular notions have a tendency to be repeated and to persist. Rice, in his *Tiger-shooting in India*, wrote that "the Kol-Bhalu is an aged, mangy, worn-out jackal, that has either left or been expelled from his pack; he devotes himself to the service of some Tiger.

It is his business to discover or give warning of the whereabouts of any stray cattle he may find that will afford his royal master a meal." This is no doubt based on some popular village tale or belief.

There are probably always jackals not far from a Tiger's kill, as they are to be found near any such remains, and it is in the vicinity of its kill that the Tiger is generally observed, especially at night when both animals are on the move; it can easily be understood that this may lead to a connection being established between the two in the popular mind as there is a connection between the peculiar cry and the presence of animals. The gathered yell of jackals is at all times piercing and like no other sound, as is the howl of a dog that lifts up its voice on hearing a bugle-call or other music. The pheall cry is still more unearthly.

The habit of the Tiger uttering a call resembling that of the sambar deer has often been referred to. The call has been mistaken for that of a sambar, and has even been said to be the ruse of the Tiger for the purpose of calling up his prospective prey, just as a man invents means to call up deer and birds. A Tigress has been seen and heard making this call so much like that of a deer as to deceive an experienced sportsman. But heard "together" the difference is discernible, and it has been said that Tiger and sambar may be heard answering one another; the sambar call has been described as "higher in pitch, more, musical, shorter, and finishing clear"; that of the Tiger "lower, more 'chesty', and not clear-cut". The call is perhaps a mate call; Mr. Dunbar Brander says that "though like a sambar bell, no sambar would mistake it."

This call is commonly known as "titting," from the Burmese term. Mr. W.S. Thom, an experienced sportsman and naturalist in Burma, says it is not necessarily either a hunting or a mating call, but always a note of alarm or apprehension uttered when the animal is suddenly disturbed or alarmed. On one occasion when a Tiger was anxious to approach a kill, but was kept off by passing villagers, it kept on "titting". He adds that even jungle people cannot always distinguish between the call of a Tiger and a sambar. It would be interesting to know whether this call is common to both sexes.

Mr. A.W. James, watching for game in the Billigirirangan Hills, heard what he thought was a sambar calling. Watching through his glasses, he saw a Tiger, saw it lower its head and its ribs contract as it made the call, and then at length recognized the high-pitched cough of that animal. It does not seem probable that the call is deliberately made to attract the deer. We know little or nothing of what passes in the minds of animals, but are prone to ascribe human modes of thought to account for their actions.

■ ■ ■

Birdwatching and Photography

By Lieut Gen Ram Kumar Gaur PVSM (Indian Army) (Retd.)

Year of Publication: 1994

Lieut Gen Ram Kumar Gaur, P.V.S.M. (Retd.) is a keen amateur photo-naturalist and anthropologist who has made good use of the unique opportunities offered by Army service in remote border regions. He has been an active pioneer in the Army's nature conservation efforts, organizing nature sutdies and encouraging his younger colleagues to participate in both adventurous and creative pursuits. He is the first person to have done a comprehensive photo-coverage of the flora and fauna of Ladakh. He is a recipient of the first *Akhil Bharatiya Bishnoi Jeev Raksha Sabha* award for conservation.

Observing birds and recording data about them is not merely a discipline of ornithology; it is also a highly rewarding hobby and leisure-time pursuit, capable of giving lifelong pleasure and deep satisfaction to amateurs as well as professionals. Over the past few decades, birdwatching has acquired immense popularity in several countries, including India, and specialized ornithological and birdwatching excursions are now offered at many places.

For an absolute novice, birdwatching can begin with things as simple as observing and identifying birds in the immediate neighbourhood and learning some interesting facts about them. This interest, once aroused, forms the very basis of serious bird study, including photodocumentation, sound recording, sketching or painting, bird counts and censuses; and to a more formal, academic or professional involvement with ornithological work.

One does not have to go far to begin birdwatching. Several species of birds will be living around the house, in the nearby garden, by a canal or pond close to your residence, and in the woodland just beyond. Birds are found even in the most crowded and noisy concrete jungles. For most beginners the only equipment needed is a pair of good binoculars, a good guidebook, and a notebook. It is extremely important, however, even for the amateur, to be sincere and systematic in his approach.

For really satisfying and systematic study, one must visit a variety of habitats within reach, periodically and regularly over the whole year, e.g. seashore cliffs, rocks, beaches, sand banks of rivers, tidal flats, marshes, lagoons, lakes, ponds, streams, scrub/thorn bushes, thickets and hedges, wastelands. grassy meadows, lone trees, groves, orchards, fruiting/flowering trees, forests, parks, gardens, lawns, farms and forested hills and bare slopes. Try to locate and record nests, perches and favourite feeding areas and times, watering points and song posts. Listen carefully, memorise and differentiate between contact calls, warning calls, alarms and songs. Quite often bird sounds help in detecting the presence of birds before sighting them visually and in many confusingly similar birds, sounds are more definitive indicators for identification.

Crested Serpent Eagle *Spilornis cheela*

One begins by becoming familiar with the area one is going to birdwatch in and with its avian diversity. Of course, on an unplanned excursion this may not be possible. For a general overview of birds in India and its various regions, it is best to read one or more of the several excellent bird books. There are also a few good magazines on Indian wildlife. Most of them devote considerable space to birds and birdwatching.

The best approach, however, is to develop an association with local, regional or national societies of naturalists. The advantages of doing this are many. Working in isolation, it is nearly impossible to crosscheck the accuracy or veracity of one's observations. There are many societies and associations doing commendable work with amateur ornithologists in India. Some of them have journals which provide a veritable goldmine of information on the birds of the region. Serious work requires access to books, journals and checklists. It is neither always feasible, nor advisable, to acquire everything personally, but one can always borrow equipment or exchange information, if one has links with local groups or individuals with a similar interest.

A few tips about the equipment and other fieldgear. Good binoculars are necessary if one intends to do something more than gaze at birds. For amateur work, magnifications of between 6 x 30 and 8 x 40 will generally suffice. More powerful binoculars, say 10 x 50 or 12 x 50, are helpful when observing smaller species from a distance. Spotting telescopes with magnifications of 16x or 20x, mounted on tripods, are necessary for those interested in studying nesting behaviour, waders, or waterbirds on distant sheets of water.

Black-necked Crane *Grus nigricollis*

One ought to wear earth colours for field work. Bright colours themselves do not necessarily alarm birds, but one's presence is given away quicker and then birds become wary, either staying in hiding or keeping a greater distance than they would normally do. It is known that two human features – the head and shoulder outline, and the bipedal stance – are alien to animals and birds and arouse instinctive fear. Therefore, serious wildlifers contrive to conceal themselves fully or at least these two features. I sometimes simply drape a green or khaki mosquito net over myself, after making a provision for limb movement.

Various types of hides have been designed and used by hunters and naturalists. The simplest is a canvas cubicle supported on sticks or a metal framework. One to three sides may have portholes with flaps for sighting. There are numerous variations on this basic design: standing hides, sitting hides, roofed or roofless hides, collapsible frame hides, zip-up hides and so on. You can prepare an inexpensive, portable hide with the help of gunny bag material, with irregular green and brown cloth patches and strips stitched on the outer side, and using four sticks. Paint, instead of cloth patches, will do equally well. Four additional cross sticks may be tied if one wants it to be a semi-permanent structure, although this reduces the hide's wrap-and-carry convenience. This type of hide is suitable for observing birds from the ground, but some sort of a *machan* is a must if one intends making observations from high up in a tree.

Siberian Cranes *Grus leucogeranus*

Whichever hide one chooses or fabricates, it must be functional, comfortable and equipped for a prolonged stay. An ideal hide is waterproof, well ventilated, roomy enough for movement, have pegs or pockets for hanging equipment, and peepholes with collapsible flaps at both sitting and standing heights, preferably in all directions. Small items such as a stool or a folding chair, a water bottle, some refreshments, a towel, mosquito repellent, etc. may appear to be luxuries, but when one actually starts

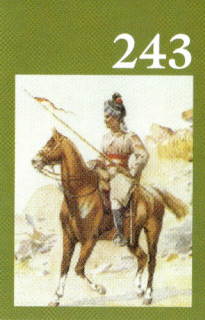

using a hide extensively, one soon realizes their worth. A closed car or jeep can make an excellent observation hide. Birds of the open country, particularly those found in the vicinity of roads with traffic, treat closed vehicles as natural objects. So long as one stays inside the vehicle without sticking out parts of the body, most birds will remain at ease and allow prolonged observation from close quarters.

However, if one cannot, or does not wish to, encumber oneself with a hide, there are simple field techniques one should employ to cause as little distraction or alarm to the birds as is possible. Break the head outline by wearing a floppy cap with a wide brim. This will also cut out the glare when looking into the bright sky, and keep the head cool in summer and warm in the winter. Needless to say, choose a dull, earth colour. Wherever possible, stand or crouch behind vegetation or a tree bole, so that the legs are hidden. Make no sudden or jerky movement, and ensure that companions observe the same rules. Talk only in low tones (whispering does not help actually, because whispers carry far). And avoid making clinking sounds if there is metal or hard plastic in the equipment. Wear the same, or similar coloured, outfit each day in the field. This helps to get the resident birds and animals of the study area accustomed to one's appearance and presence. Let them judge the watcher's harmless intentions, by not making an early attempt at approaching them, their nests or the young too close. In any case, one must not touch or otherwise disturb bird nests, eggs or young.

Great Indian Bustard *Ardeotis nigriceps*

If one chances upon a baby bird that has dropped out of its nest, one should simply put it back in place if one can locate and reach the nest. Contrary to popular belief, parent birds do not reject their young just because they have been handled by humans. If one cannot locate or reach the nest, one should place the baby carefully on a branch or the top of a shrub beyond the reach of a cat or a dog its hunger calls will soon attract the parents and they will feed and guard it. It is extremely difficult to ensure the survival of unfledged chicks if they are brought home, since the nature of their food, the frequency and intensity of feeding, their hygiene requirements and, above all, their need for parental warmth and brooding, are nearly impossible to replicate. Faced with such a situation, the best course is to overcome emotion and contact one's birdwatching circle for help in locating a suitable facility–a bird hospital or an experienced birdkeeper. Till such time as the young bird is in safe hands, one may have to keep it warm, dry and frequently fed on juicy worms, as these are a safe bet for most bird babies; any trials with other food may kill the bird, unless one knows its diet and feeding method.

Early morning is the best time for observing birds, as it is the peak of activity and vocalization for most of the species living on the ground, in shrubbery and in trees. Some of the night foragers can also be seen returning to their resting places in the morning. Although in wooded areas the raptors are not hunting in the early morning–their feathers are usually a little damp from the dew of the previous night–they can be seen fairly frequently, perched high on treetops or poles, trying to catch the first rays of the sun to get crisp, while they scan the countryside for potential prey. In fact, one should take advantage of such activity patterns of birds to schedule the day and cover as many habitats as possible. Use the early morning to go to

a wetland or marsh where the birds will be active, move to the surrounding woodland as the sun comes up and use the later part of the morning to watch the day feeders out in the fields. In coastal areas and on the larger water sheets, bird activity continues nearly all day long. Late afternoon and evening are ideal for observing crepuscular species such as night herons, owls and nightjars, and to observe the daytime feeders, such as cranes and some ducks and geese, returning after feeding in the fields.

In the West, birdwatchers frequently use tape recordings of bird sounds to attract birds in the field and also for making comparisons with their observations. Some authors suggest the use of imitation of bird sounds or mimicked distress calls to attract birds. In India, however, I am not aware of anyone having made serious use of either method for bird study. Tape recordings of Indian birds are not available in India as they are in some Western countries, and the average Indian birdwatcher does not have access to the gadgetry needed for such recordings. As to the second method of mimicked distress calls, its efficacy is proven from the good old days of duck shoots; you only have to ask an experienced hunter, or in our age, a poacher, about it.

It is extremely important to gather data in a thorough way, adopting wherever possible standard terms and units and standardized methodologies, if you are aware of them. I have found the following standard form devised by the British Trust for Ornithology quite useful.

Black-crowned Night heron
Nycticorax nycticorax

1.   Name of birdwatcher.
2.   Name of any birdwatcher accompanying you on the sortie.
3.   Site or location of observations–refer to well known landmarks.
4.   Date of observation–mention times at which observations were made and the percentage of the total area covered during the sortie.
5.   Data gathered: type of vegetation, weather conditions–cloud cover, temperature, wind, rain, visibility, etc.; environmental features–type of terrain: rock, stony, gravel or scree, coarse sand, fine sand, mud or ooze, clay, humus, peat, etc; sloping or otherwise, presence of surface water expanses or other water source, man-made buildings or other structures nearby, etc.

Painted Stork *Mycteria leucocephala*

Purple Sunbird *Cinnyris asiaticus*

6. List of species observed–for each species indicate the number of different sightings made; if possible make notes of the sex of the individuals sighted, the sighting with the largest number of individuals observed together and the approximate number of individuals seen or present in the area; all the various observations about behaviour and other features that you consider worth mentioning.

For identification the following points need careful observation and recording:

(a) Size–comparing with the more common species like sparrow, myna, crow, domestic fowl, geese, sarus.

(b) Colour–Overall coloration, associated colours, colour of pupil, any noteworthy pattern/ patch/streaks/bars/spots on head, neck, breast, cheeks, wings, tail, beak, legs, and around the eyes, and moustachial streaks. Some colour patterns and other diagnostic features are visible only in flight.

(c) Shape and size of head, beak, tail, wings, legs, claws, spurs, crest, webbed feet, cover on legs.

(d) Behaviour–shy, bold, tame, neurotic, silent, strident, fidgety, flitting, gregarious, loner, pairs, groups, flocks, pugnacious, sexes in mixed groups or isolated.

(e) Flight pattern–straight, undulating, take off and landing, zigzagging, spiralling, wing beating pattern and frequency.

(f) Surface movement–hopping, walking, running, waddling, swimming, dabbling, diving.

(g) Feeding method and food.

(h) Perching posture.

(i) Nest shape, material, location, construction, egg colour/pattern, size, numbers.

Such record keeping organizes one's data into a systematic format, makes it immediately utilizable, and comparable with available standard information. Note down what is observed, making definitive statements only when sure, keeping queries for later clarification by a more experienced birdwatcher or with the help of a manual, Use the

European Golden Oriole *Oriolus oriolus*

concept of elimination to narrow down the identification of species. Beginning with a number of probables on the basis of general similarities, reject those that do not precisely match the description personally noted, with the help of the identification key in a field guide or handbook. Besides physical descriptions, use other observations such as location, season and time of sighting, any salient behavioural aspects, etc. to arrive at the correct identification. In the beginning one need not delve beyond recognizing the gross characteristics of Orders and Families. Use one or more authentic handbook or guide for identification. But a word of caution here. Confusion frequently results when one cross refers two guides, because these can differ on basic descriptions, and colour sketches actually do not agree for many species. A number of Indian productions suffer from inaccurate colour reproduction and poor drawing quality.

Photography

Bird photography is a very challenging field and requires adequate knowledge of bird behaviour as well as of some special techniques of photography. It is not within the scope of this book to teach the basics of photography. I wish, instead, to share with the reader some general information and some ideas based on my personal experience, accumulated over two decades of nature photography.

For reasons of economy, versatility and portability, the 35 mm single lens reflex (SLR) is the most suitable camera format for the wildlife photographer. Medium and large format cameras and lenses are too heavy for the field, besides their high costs and the limited range of optics and accessories. Autofocus and programmed cameras do have many advantages in some fast action situations. The auto exposure (AE) mode is certainly helpful in as much as it cuts down on the time taken in estimating exposure and reduces missed shots. But in the unusual light and shade conditions encountered in bird habitats–dark object against the sky or against a bright background, dappled light through foliage or a dark background with a small main object illuminated by angled light, and so on, the simple auto-exposure systems tend to go wrong and require manual corrections based on the photographer's experience. More advanced multi-pattern-segment (MPS) auto-exposure cameras are better equipped to deal with such situations.

Zoom lenses offer a compromise solution of the need to have several focal lengths to cover all distances, situations and bird sizes. Wide-angle to normal focal lengths (28-85 mm or 3570 mm) are ideal for general habitat, large flocks and sunrise/sunset shots; short to medium telephoto/zoom lenses (60-300 mm, 70-200 mm, 80-210 mm, or single focal lengths from this range) for large birds at moderate distances; and long telephoto lenses (400 mm upwards, to possibly 1000 mm) are essential for small birds and long-distance coverage. There is a whole new range of compact telephoto lenses with mirror optics available these days, but they have severely limited depths of field and no aperture adjustment. As a result one has limited control over exposure time. Tele-extenders or converters boost the focal length to give a larger image size, but the resultant losses in sharpness and brightness are serious handicaps. If one can find a 1.5x or 2x converter with reasonably good optics (sharpness, flatness of field), one should be able to put it to good use, at least in brightly lit situations.

A strong flash is useful in lighting nest holes and birds in deep shade or foliage, or as fill-in. A sturdy tripod is a must for those wishing to set up at one spot for long sessions. Its use is recommended while shooting with long lenses, to avoid pictures with blur. A good tripod should have well placed, smooth operating controls for elevation, tilting, swivelling and locking. It should be able to take the combined weight of camera body, long lens, motor drive and tele-converter, without wobbling. But if a tripod is not available, one can still hand-hold long lenses by perfecting one's breathing and camera holding techniques to minimize camera shake. One can use a rigid support–tree trunk, wall, car or box–to brace oneself against or to place the camera on. Practice the postures, grip and shutter release techniques recommended by the camera manual and photographic guides. Do not shoot from inside a vehicle if the engine is running. Even otherwise the vehicle body is unstable, particularly if there are other occupants in it.

If possible keep an additional manual camera body, which takes the same range of lenses and accessories, but is free of the shortcomings of the auto/electronic (jamming on battery failure, exposure goof-ups in tricky lighting situations, vulnerability to vibrations, heat, cold, etc.). Motor drives or autowinders have a definite advantage in fast action situations and in remote control photography. Similarly, remote cables, sensors and radio/infrared beams and photoelectric trigger devices are extremely useful in the photo documentation of fast action, flight, or where the photographer cannot position or hide himself. The list of add-ons is long but while each of these has a definite advantage to offer, one must decide what to have, based on one's requirement, intended use of the photographic output, and budget.

Among other fieldgear needed is a camera bag that is waterproof, dustproof, cushioned, lockable, crushproof, designed for easy access to its contents and preferably an earth colour. One should also carry a cleaning kit consisting of a soft blower brush, or lens tissue, a clean lint-free wiping cloth and some lens cleaning fluid. Always keep a spare set of batteries and enough extra film, because there are no replacements out in the field.

Black-headed ibis and Spoonbill

One can use black and white, colour negative, or colour transparency film. The choice will be entirely governed by one's personal liking for a particular medium (for instance some photographers find black and white more exciting) and the ultimate use one intends to put the pictures to. If the intention is to use them for presentations, publication or educational work, then the best medium is colour slide film. But if one desires to build up personal albums or portfolios, one may prefer colour or black and white negative film, from which enlargements can be made to specifications. The choice of film speeds depends on the nature of work. For general habitat or bird pictures, using the camera on a tripod in bright daylight, film speeds of 50, 64 or 100 ASA/ISO are very good. To freeze fast action as well as for hand-held long lenses, or for dim light situations, faster film speeds of 160, 200 or 400 ASA/ISO are required. Of course, the fineness of picture grain and therefore reproduction quality–declines as film speed increases. So one has to strike a balance, using one's own judgement.

Now for a few practical hints. Check your equipment periodically, especially for signs of attack by dust, moisture, grit, fungus or battery fluid. If your equipment travels a lot with you, check occasionally for screws and other fasteners getting loose by vibration, and tighten them gently. You may apply small drops of non-corrosive adhesive (or nail lacquer) on the exposed screwheads which will prevent the screws from coming loose. Protect your equipment from motor vibration and bumps by cushioning various items with pieces of sponge or bubble-packaging material. Carry it on your lap or on a well padded seat during vehicle rides; never place camera bags on the floor of the aircraft or vehicle or in the luggage boot. Do not place film or equipment in the glove compartment or on the dashboard of a car because these heat up excessively. In the field, store all your gear in the shade and out of harm's way. Whenever handling a camera, make it a habit to first put the strap around your neck. If you ever happen to drop it, this will prevent it from falling to the ground.

Darter *Anhinga melanogaster*

The selection of subject matter is essentially a personal choice, at least for amateurs. Some are turned on by the ferocity and power of raptors, others by colour and patterns of plumage, still others seek rare or endangered species. But most prefer general photocoverage according to opportunity and availability of the subjects. Having once decided on the subject, study guidebooks and other literature to know the when (season, time) and where (location, habitat) of the selected species. These are the same factors to bear in mind as discussed earlier for birdwatching.

Observe the activities of the subjects to establish patterns based on their habits, routines and natural cycles. Learn to anticipate events and movements of birds, instead of chasing them all over for 'pot shots'. Practise pre-focusing at spots where one expects the birds to arrive. For example, a woodpecker will usually climb a tree in spirals, so focus where it can be expected to emerge next from behind the tree trunk; similarly, locate favoured perches, songposts, trees in fruit, flowers with nectar, entrances to nest holes, water points (especially in and regions) and so on.

Nondescript pictures have little use; they waste photographic material and block money. So learn to be selective and critical in your shooting. If you are doing portraits, snap when the bird has a pleasing posture. Choose an aperture for a depth of field adequate to get at least the bird's head and eyes (and preferably the diagnostic features as well) in sharp focus. Eyes should be well lit, so that they have what, photo editors call 'catchlight'. In action photography, press the trigger a fraction of a second before the peak action, so that the shot can actually capture that moment. Use flash to fill in and soften harsh shadows; do not try to eliminate shadows altogether or the modelling effect will be lost. Take help from nature photography guides and existing picture portfolios of known bird photographers to develop an aesthetic sense of composition and framing. if planning to put the pictures to commercial use, practice bracketing of important shots, i.e. shooting a graded sequence of exposures, various depths of field or bird postures, so that a selector may exercise choice.

There is a great need for accurate photographic documentation of Indian birds, both for publication and for the scientific study of all aspects of bird life: nest building, incubation, feeding, flight, courtship, populations, threats to survival, conservation status, and so on. The whole world of birds is waiting out there, to be observed, photographed and documented. So have a field day birdwatching and shooting and, while you are at it, have fun!

Oriental White-Eye *Zosterops palpebrosus*

Wild Life Preservation: India's Vanishing Asset

By Lieut Col R W Burton, (Indian Army)

Year of publication: **1948**

Born in 1868, **Richard Burton**, the sixth son of Gen. E.F. Burton of the Madras Staff Corps, was commissioned from Sandhurst in 1889. Posted to the Indian Army in 1890, he was permanently crippled by a riding accident in 1903. He was, thereafter, assigned to the Cantonment Magistrates Department. A fearless sportsman and a keen fisherman, he wrote over 200 articles on various aspects of natural history and was the first naturalist to campaign for the preservation of Indian wildlife. Col Burton passed away at his residence at Surrey, England in January 1963 at the age of 95.

THE BOMBAY NATURAL HISTORY SOCIETY

For many years the Society, through the medium of its *Journal* and other attractive publications, has endeavoured to create and stimulate in India an interest in the wild life of the country. During the past sixty years there have appeared in the *Journal* upwards of fifty longer and shorter articles and editorials on the subject. It was to a great extent owing to the Society that Act XX of 1887, 'An Act for the Preservation of Wild Birds and Game' (passed after nearly 30 years' agitation in the matter), was replaced by 'The Wild Birds and Animals Protection Act (VIII of 1912) which, together with the Indian Forest Act (XIV of 1927), is the basis of all rules in force at the present time.

PRINCIPLES

In all civilized countries there is a general recognition of the need for concerted and practical measures to stop the forces of destruction which threaten wild life in all parts of the world. The principle is the same everywhere, the methods to be employed must vary in every country, and will also vary in different parts of the same country. That has special application to India as a whole, and is the reason why legislation on wild life in this country has been complex and difficult.

'Until it is recognized that Wild Life is a valuable natural resource, and the benefits derived from an unguarded resource are wasting benefits, waste will continue until the resource has gone and the benefits have vanished. No natural resource is more sensitive to conservation than Wild Life, and no natural resource has suffered more from lack of conservation. During the last sixty years species have been exterminated due to this deficiency.' (Hubback).*

*Principles of Wildlife Preservation. JBNHS 40: 100 (1938)

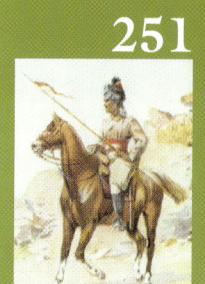

At the present time the pace and extent of the waste is alarming. In this country there is the gravest need for concerted action.

'In its fauna and flora nature has endowed India with a magnificent asset. A further interest attaches to our wild life from its association with the folk-lore and legendary beliefs of the country. It is an interest not confined to India alone, but which has spread among men of culture everywhere because of the esteem and admiration in which her sacred books and writings are held.' (Prater).*

BIRDS

Although birds are not now persecuted to the same extent as animals, yet an enormous amount of unnecessary and preventible damage is going on. One bright spot in India, as Champion has remarked, is that non-game birds are not harried to the same extent as used to be the case in some western countries, for the Indian boy does not amuse himself by uselessly collecting vast numbers of birds' eggs. But India had the dreadful plumage trade, which was far worse.

The Great Indian Bustard is becoming increasingly scarce and has gone from areas where it was common not many years ago. The Monal Pheasant and the Tragopan of the Himalayas have been saved only through prohibition of export of plumage. Other birds saved from what would have practically become extermination through the extremely lucrative plumage trade were peacocks and black partridges, egrets, junglecocks, paddy-birds, kingfishers, jays and rollers, orioles and a host of others. The governments controlling Pondicherry, Goa and other ports on the coasts of India cooperated, so the traffic was stopped. But there were many subsequent cases of smuggling, and these will certainly recur if the plumage trade measures are ever relaxed.

Great Indian Bustard *Ardeotis nigriceps*

Nomadic Tribes

In all tracts where the snaring and netting of ground game is the hereditary occupation of various nomadic tribes, partridges, quail, florican, hares are fast disappearing. These people, expert in their calling for untold generations, sweep the country as a broom sweeps the floor; nothing is passed over, nothing is spared.

The time has long past when snarers of indigenous game birds should be allowed to continue to earn a livelihood in that way; in any case all markets should be denied them, and public opinion should recognize that flesh of such wild creatures is not in these days at all necessary for human existence and should ban the killing of them for food alone.

Within a considerable distance also of Calcutta, Bombay, Madras and other large places markets are supplied in season and out of season through other agencies and local 'shikaris' in spite of the local government close season rules under Act VIII of 1912.

Even if legislation on lines suggested in this article is effected, however much it can and may help in the endeavour to protect wild life, anything like *practical* success is possible only if there is a strong Public Opinion cooperating with the governments. That cannot be too often reiterated.

Himalayan or Monal Pheasant
Lophophorus impejanus

Value of Birds

In connexion with all that is written above the thought-provoking article, 'Bird Protection in India: Why it is necessary and How it should be controlled', by Sálim A. Ali, M.B.O.U., contributed in 1933 to the U.P. Association, should be read by all governing bodies. Indeed it is most essential to national India that bird life should be adequately conserved. For 'Quite apart from a sentimental value, birds render incalculable service to

man. Without their protection our crops, our orchards, our food supply would be devoured by hordes of ravaging insects. Birds are the principal agency that controls the bewildering multiplication of insect life which, if unchecked, would overwhelm all life on this planet.' (Prater).*

SPECIES IN DANGER

Mammals: The Great One-horned Rhinoceros has only been saved by special measures and these, if in any way relaxed, will inevitably lead to its extinction. A close relative to the above the Lesser One horned Rhinoceros (*R. sondaicus*), which has been within the memory of many an inhabitant of the Sunderbans jungles and other tracts, has completely disappeared–none now exist on the soil of India. The Asiatic Two-horned Rhinoceros which occurred in parts of Assam has gone from there for ever, and both these species are approaching the vanishing point in Burma and other countries where they were formerly in fair number. In Burma the Thamin Deer is probably doomed to extinction.

Asiatic Twohorned Rhinoceros *Didermocerus sumatrensis*

In Western Pakistan and neighbouring mountains the Straight-horned Markhor is rapidly disappearing; and if the Punjab Urial is not carefully preserved that species will not long survive.

The Indian Antelope (Black Buck) is becoming increasingly scarce and will eventually only be preserved through protection; to a less extent the same can be said of the Indian Gazelle. The Cheetah or Hunting Leopard, was not uncommon in the central parts of the peninsula but is now practically extinct in a wild state. The Wild Buffalo has almost gone from the areas east of the Godavari river where it was common not long ago; and it needs continued protection in Assam. The Asiatic Lion in India has only survived in its last stronghold through protection in the Gir Forest of the Junagadh State in Kathiawar.

'In many districts the larger animals have been totally wiped out. In others, where they were once common, they are now hopelessly depleted. There are a few parts of India where the position of wild life is to some extent satisfactory, though insecure. Equally there are extensive areas where conditions are so appalling that, if left unchecked, they must lead to the complete destruction of all the larger wild creatures which live in them.' (Prater *op.cit.*)

Year in year out there is terrible destruction throughout the enormous tract of mostly hilly and forested country comprising the Eastern States, from the Godavari river as far as Bengal, some of which are being now merged into India. The methods of the aboriginal tribes inhabiting

Wild Buffalo (Bull) *Bubalus bubalis*

this huge area (and other parts of India also) are those of extinction, for they net, snare, shoot all edible living creatures at all possible seasons and particularly during the hot weather months when water at the few pools is a necessity to all and renders them an easy prey.

In the Himalayan mountains also where control is difficult wild animals are definitely decreasing, and only to be found in any number in the more inaccessible places.

Time for Decision

The Governments have to decide without delay if wild life is to be effectually preserved or the present lamentable state of affairs allowed to continue. In the latter event there can be but one result–the total and irreplaceable extinction of some forms of wild life, with everywhere woeful reduction in number of all wild animals and many species of birds.

There is no middle course. Half measures will be futile and waste of time. Wild life is a national and natural asset which, if it is ever lost, can never be replaced. It is necessary that governments should give a lead, a strong and unambiguous lead.

India and Pakistan should be proud to stand side by side with other civilized countries of the world in saving their fauna from extinction.

In these days public opinion should recognize that flesh of wild animals is not necessary to human existence; but public opinion may not eventuate for many a long day. Meat-eaters want something for nothing and care not how they get it. Posterity means nothing to them.

One instance. In November 1947 six shot carcasses of chital hinds were found with a man in a country bazaar. Police said prosecution was doubtful because there was no evidence as to where the animals were killed. But a so-called 'Sanctuary' was not far distant. The burden of proof should be on the possessor. In any case Rules under Act VIII of 1912 must have been contravened and conviction could have been had.

Legislation, and that very speedily, should absolutely prohibit offering for sale, possession for sale, or marketing in any way the hides, horns, flesh or any other part of any indigenous wild animal throughout the year. And, as was done by Notification in 1907 to suppress the plumage trade, so also should the trade in products of wild animals be stopped by prohibition of export by sea, and by land now that Burma is independent of India.

It would appear that there is no possible objection on religious or other grounds to a general law throughout the whole country to the above effect. Profits are large and really deterrent sentences would be necessary.

Public Opinion

At the present time public opinion as to wild life preservation is almost non-existent in this country. It is only through public opinion that wild life can be saved and preserved through all the future years.

Hear a great statesman of former days in another land:

'In proportion as the structure of a Government gives force to public opinion, it is essential that public opinion should be enlightened.' (George Washington)

In these days that is done through the many avenues of propaganda.

Propaganda

Political parties in this country have been able to rapidly rouse and educate public opinion in all kinds of political matters. Is it not therefore possible that through like efforts the thoughts of the people can be directed towards the necessity of the conservation of this national asset of wild life? It *can* be done. Great are the powers of propaganda.

Let the secular and religious leaders of the people lend their great influence and abundant powers of persuasion to furtherance of this most pressing need. Where there are religious influences at work wild life is sacrosanct. All of us know that. We see it everywhere: peafowl, parrots, pigeons, monkeys, nilgai and other species–fish in sacred river pools. Even where some of these creatures cause much loss and damage to growing crops and to gathered grain they are protected through religion. Where the rulers of States have made their wishes known and enforced, wonderful are the results. Instances are known of those having this subject at heart; a number are known to the writer.

ENEMIES TO WILD LIFE PRESERVATION: FORGETFULNESS-INDIFFERENCE-IGNORANCE-GREED FOR GAIN

Laws are enacted, rules are made and forgotten, for there is no continuity of official enforcement and no public opinion to keep them in mind.

India

Within not many years Act VIII of 1912 was forgotten, the wide scope of its provisions unknown. The rules under the Act were ignored and its provisions a dead letter.

1933. 'The Governor in Council has reason to believe that there has been little improvement in the administration of this Act and that subordinate officials are, not infrequently, offenders against -its provisions, x x x x it is believed that sheer ignorance of close seasons is in many cases the cause of offences against the Act ....

Guns and Greed

It is the possessors of guns and rifles who do the greatest amount of harm. In many cases it is not the actual licensee who does the damage, but the illegal habit of lending or hiring out the weapon to others. Could the abuse of licence granted as a personal privilege be stopped much good would result. But how is this to be done? Only through public opinion could it be effectually curtailed. So what ? Suggestions as to Arms Act, if carried out, would do some good.

It is as a poacher that man is the great destroyer; and the main incentive is profit by selling hides, horns, meat–to a less degree, is it meat only. In some places local dealers finance the village shikari, providing him with guns and ammunition in exchange for hides, etc. Sambar and chital hides and heads are openly bought and sold in many bazaars and there is nothing to prevent it. Sale of trophies is common in many large towns and cities. To deprive sellers of their markets by effectively enforced legislation and through public opinion is the only way to remove temptation to kill for profit. If there were no buyers there could be no sellers. Utopia!

CROP PROTECTION

It has always been pointed out, and is notorious, that crop protection and other weapons are used for the slaughter of game animals in adjacent and further forests regardless of all laws, rules, age, sex, season, or any other consideration whatever than profit. All this and other poaching is mostly carried on in Government forests, for there are to be found more animals than outside them. So far as crop protection goes the argument in the mind of the cultivator is that if there are no animals the crops will not be eaten, so he may as well hasten the coming of the welcome day and meanwhile make money for himself and provide meat to the community.

Guns

The great increase over former years in the number of licensed guns is producing its inevitable adverse effect; and there is the mass of unlicensed weapons carefully concealed and constantly used. While the reduction in the number of weapons is admittedly a difficult matter, the withdrawal of crop-protection guns during the seasons when the crops are off the ground and the guns not needed for legitimate use is a reasonable proposition. That would be of much benefit as those are the months in which they do the most harm.

A suggestion from Assam was that crop-protection guns now owned by villagers (more especially those inside Government forests) might be acquired by Government for temporary issue at the right time and withdrawal when no crops, or for other reason.

It is not likely, however, that Provincial Governments would adopt these gun withdrawal suggestions on account of practical difficulties and extra work to District Magistrates and other officials.

A proposal advocated by many is that crop-protection weapons should be licenced for cut-short barrels only. Cogent arguments against such modified weapons are that they are more liable to be loaded with buckshot, so causing many animals to be wounded and lost; are dangerous in hands of such persons as ordinary cultivators; and such restriction would cause an increase of concealed weapons for poaching.

It has been demonstrated in South India by Colonel R.C. Morris that bamboo-tube rocket-firing 'guns' are both cheap and effective for scaring crop-raiding wild elephants, so firearms need not be used against them by cultivators.

Such 'guns' could also be effectually used in many forest areas against other crop-raiding animals and so enable a large reduction in the number of guns now licensed for ostensible crop protection.

Jerdon's Courser
Rhinoptilus bitorquatus

AGRICULTURE AND WILD LIFE

For purposes of wild life conservation lands may be classified in five main categories: *Urban-Agricultural-Waste-Private -Forest.*

Urban Lands

In these, measures should be taken for the protection of all birds. Areas actually under the control of municipalities or local boards could with advantage be constituted bird sanctuaries where the killing of, or taking the eggs of, any wild bird should be forbidden.

Agricultural Lands

Here lies the clash between the interests of Man and Animal; for which there are two main reasons.

Firstly, the population of the country is increasing by about five millions yearly, so the areas under cultivation are extending, and must continually extend to the utmost limit, which means the continual absorption of all cultivable waste lands and secondary forest lands.

Secondly there is the imperative need of protecting present and future cultivated lands from wild animals.

In some parts, where cultivation is contiguous to or near Reserved forests the depredations of wild animals present one of the most serious handicaps the cultivator has to face. The animals are not only deer and pig and some species of birds, but nilgai, monkeys and parrots which are protected by religious beliefs.

'Human progress must continue, and in the clash of interests between Man and the Animals human effort must not suffer. But this problem has been faced by other countries. Cannot a reasonable effort be made to face it in our own? That an intensive development of the agricultural resources of a country may accompany a sane and adequate policy for the conservation of its wild life is shown by the measures taken to this end by all progressive countries.' (Prater *op.cit.*)

Pinkheaded Duck
Rhodonessa caryophyllacea
(Extinct)

But in those countries there is universal literacy, a people easily educated to a proper public opinion, and where the masses do not clamour for possession of guns and rifles and even for repeal of the Arms Act.

Waste Lands

These are beyond redemption as to wild life, and in any case all that are at all cultivable will soon be merged with Agricultural lands.

Private Lands

The general concensus of opinion is that in most ordinary tracts the position is hopeless. The people have been educated to destroy, and there is no agency to stop it. Only through the owners themselves and through propaganda can any change be wrought: and before these operate the position is likely to be beyond any remedy.

Some-private lands, however, have forests for which rules have been framed to regulate hunting and shooting, while in others no rules have been framed. The wild life situation in all these depends on the amount of control exercised by the landowners. Some United Provinces landowners maintain renowned Swamp Deer preserves.

The Wild Birds and Animals Protection Act, 1912, deals with the right of private owners only in so far as it prohibits the shooting of the specified animals whether on private lands or elsewhere. This prohibits private owners killing females of deer, etc., and killing during prescribed close seasons.

Government Forests

These are of several kinds and mostly under the Forest Department, but some are under Revenue Department; none of the latter are Reserved Forests.

While it is essential that the cultivator should have reasonable latitude to defend his property, it is equally essential that there should be certain areas of Reserved Forests, where the laws and rules for protection of wild life are, or should be, rigidly enforced.

State-owned Reserved Forests, similar forests in the Indian states in the Terai tracts of the frontier states of Nepal, Sikkim, Bhutan and some private forests are now, and must continue in future to remain, the natural sanctuaries for wild life in this country.

Wild Life and / or Game Associations

Where these exist they are, if well organized and conducted, wholly productive of good. There is the Association for the Preservation of Game in the United Provinces through which the All-India Conference for the Protection of Wild Life was held at Delhi in January 1935 and the Hailey National Park established in the Kalagarh Forest Division. At the Conference it was declared that, "Indian Wild Life could only be saved by Public Opinion, and that legislation, however efficient, could do little in matters like these without the whole-hearted support of the Public." How true. Where is the Public Opinion? Where is the support of the Public? What is the state of wild life at this thirteen years later date?

Swamp Deer
Cervus duvauceli

In Northern Bengal are three shooting and fishing associations:

(1) Darjeeling Fishing and Shooting Association.
(2) Tista-Torsa Game and Fishing Association.
(3) Torsa-Sankos Game and Fishing Association.

In Madras is the 69 years old Nilgiri Game Association but for which little wild life would now exist in that district. Continuity of purpose, efficient control.

In 1933 an Association for the Preservation of Wild Life in South India was inaugurated at Madras by the then Governor of the Presidency, but it came to nothing and has never been heard of since then. Continuity of Purpose—Public Opinion these basic essentials do not exist. Without them there can be no effectual preservation of wild life for posterity.

NATURAL ENEMIES OF WILD LIFE

Tigers. Where in forest areas deer have been excessively reduced in number through poaching the tiger turns increased attention to cattle killing. The tiger needs the pursuit of deer to satisfy his hunting instincts, and where the balance of nature in this respect is not unduly disturbed he is of benefit, as also the panther, to the cultivator of land within the forests and along its borders, for he keeps the deer and wild pig population within natural limits. But where the stock of deer is unduly reduced not only are all the deer killed out but the tiger is forced to prey on cattle; and as these are penned at night he is compelled to change his habits and hunt by day. That is when he takes great toll of grazing cattle and sometimes turns against the people also. Then the cultivators clamour for protection from the menace brought about by the unlawful poaching done by themselves and others.

Panthers. These are less destructive to village cattle as they prey on sounders or pig and a variety of smaller animals ordinarily ignored by the tiger; but they also kill cattle and other domestic stock to a greater extent when the balance of nature has been disturbed. In areas where panthers have been unduly reduced through rewards for their destruction there has resulted such an increase of wild pig as to necessitate rewards to reduce their number.

Predatory animals have a distinct value as a controlling influence against over-population by species whose unrestricted increase would adversely affect the interests of man.

UNNATURAL ENEMIES

Cattle diseases. A great cause of much periodical mortality to buffalo and bison is through rinderpest. Against the introduction of this by grazing cattle effective action has been found impossible.

CROP ENEMIES

Elephants. Effective legislation was enacted in 1873 and 1879 to protect the elephant. In these days of mechanical haulage these animals are not necessary in such large number as formerly.

It can be anticipated that the demand for elephants in India will before long be reduced to the few needed for timber extraction in difficult areas, for riding and transport duties by the Forest and Wild Life Departments, for ceremonial purposes, and for zoological purposes.

Wild Pigs. Deer and the like are crop raiders, but it is the wild pig which is the principal crop destroyer both in the open country adjacent to the forests and within the forests. Where the balance of nature has not been disturbed the larger carnivores take care of the surplus pig population harbouring in the forests. It is not by the lone working cultivator with his gun that any impression is made on the number of pig.

Some 25 years ago it was realized by the Bombay Government that damage to crops by wild pig amounted to crores of rupees. Measures to deal with the trouble outside reserved forests included

NIKHIL BHOPALE

Wild Boar *Sus scrofa*

clearance of cactus and thorn thickets and other such coverts together with organization of inter-village pig drives. Those measures will have had good results if continued as a fixed policy, but not otherwise.

At the time of writing (end of January 1948) the Government of India have been asked by the Government of the Central Provinces to supply arms and ammunition for use of cultivators against wild pig. If the weapons are used against pigs only, and at organized drives only, good may result, but if not so controlled they will assuredly be turned against the fast dwindling wild life.

NATIONAL PARKS

Those who have knowledge of the subject are of the opinion that India is not yet ready for these. The Hailey National Park, the situation of which conforms in most respects to conditions laid down for a sanctuary is specially situated and may be a success. A full account of it would be welcomed by members of the Society.

The Banjar Valley Reserved Forests area in the Central Provinces is perhaps suited for eventual status of a National Sanctuary (not Park). The case for it is outlined by Dunbar Brander. Buffalo, lost to it not many years ago could be reintroduced; otherwise it contains all the wild animals of the plains except elephant, lion, and gazelle. Elephants are not wanted as there are plenty in other provinces.

Even fifteen years ago the area was admittedly tremendously poached.

SANCTUARIES

All sportsmen are agreed that these are of little use unless adequately guarded and, as that has not yet been found possible in India, such areas merely become happy hunting grounds for poachers from far and near. The constant presence of sportsmen of the right kind has been found the best guarantee for preservation of wild life in Reserved Forests.

There are, however, tracts and forests where wise forethought and administration can, with the willing cooperation of the people if that can be obtained, do much to preserve wild life for posterity.

Under the present reorganization of India a number of the smaller States, and many lands privately owned, within which wild life has had no regulated protection, will now be brought within the laws of the rest of the country to the benefit of wild life in all its aspects–if the laws are properly enforced.

The notable contributions on the Problem of Wild Life Preservation by Mr. S. H. Prater, the Society's Curator, on the 10th August 1933, and by forest officers for India,, and Smith and Hubback for Burma are of the greatest value and recommended for careful study by all governments in this country.

A WILD LIFE DEPARTMENT

Forest Officers of the regime now ending have been of the opinion that animals inside reserved forests should not be removed from the protection of the Forest Department and placed in the charge of a separate department. Their argument has been that the present system has worked well; such action would create resentment and alienate the all-important sympathy of the powerful Forest Department; and that a Game Department would be in no better case than the Forest Department for dealing with breaches of laws and rules.

On the other hand sportsmen and others with many years of experience are of the opinion that under the present changed conditions forest officers, while not relieved of all responsibility, should be relieved of their present whole-time onus and share the burden of preservation of wild life with a specially organized Wild Life Department.

Himalayan Quail *Ophrysia superciliosa*
(Extinct)

Red Panda *Ailurus fulgens*

Why should not the two departments work amicably in liaison?

There need be no friction. The appointment of Honorary Wardens has not always proved a success, not on account of any disagreements but because the conservation of wild life is a whole-time duty which no man with other interests and work to do can efficiently perform. There could be Honorary Wardens to assist the Government Wardens and enthusiasts could be found for that work.

It has not been that all forest officers have been keen on the preservation of the larger game animals; some sylviculturists have expressed definite opinions against any deer being allowed in the forests, but movable fencing has been found a sufficient protection to special plantations.

In these days of intensive exploitation of timber and forest produce the work of forest administration has become more and more exacting and the officers find it exceedingly difficult to give time in office and out of doors to work which brings in no revenue and is considered of subsidiary importance.

Would not Forest Officers welcome the considerable measure of relief which the formation of a Wild Life Department would afford them? Surely they would. Neither their pay nor their prestige would be in any way affected.

It has been experienced that an unbribable staff of Game Watchers has been difficult to procure. That again is strong reason why there should be whole-time Wardens whose interest would be to prevent malpractices. A Wild Life Department means that continuity of purpose without which all endeavour is of no avail.

PROPAGANDA

During the years 1932 to 1936 there was a good deal of wild life propaganda in the public press at the instance of the U.P. Association, also in South India, but all that quickly died down. Then came the war years and now the present difficult times. Wild life has greatly suffered.

Educative propaganda needs constant reminders and exhortations to the public. Only if the subject is frequently repeated will it gain a hearing.

It is commonly said that it will take years and years to arouse public opinion as to wild life. But we daily see what the present leaders of public opinion in this country can do in many ways vitally affecting the present and future lives of the masses, how speedily laws are enacted and far-reaching measures put into motion. There is, for instance, the vast organization for further education of the literates and the initiation of universal literacy for the masses. There seems to be no reason why wild life preservation could not also be given the highest priority. Some of the reforms could wait, not that they should, far from it, but the wild creatures cannot wait–and survive.

Wild life preservation does not only mean the protection of animals and birds, it means a fight against the destruction which is going on at an increasing pace–particularly against deer–and is not of Nature's ordering. It is simply asserting the right to live of the undomesticated animals and indigenous birds.

An atmosphere of mistrust and suspicion is all too common among uneducated people, so the beneficial intentions of measures towards wild life preservation are apt to be misconstrued unless the objects and reasons receive the widest publicity through Government channels, and the newspapers.

The years are passing; this great national asset is wasting away. It is the duty of every government to preserve it for posterity. The urge should come from the highest levels.

PROPAGANDA METHODS

The time is now.

The Ministry of Information and Broadcasting could make it a routine matter to keep this subject constantly before all classes of the people. Special talks could be given on All-India Radio, and other systems.

The Educational Department could cause all governing bodies, and educational institutions to issue pamphlets, organize lectures, lantern slide talks, and issue of suitable leaflets to all colleges, schools, primary schools. All this could be worked out on the lines of the anti-malarial campaign which was an India-wide effort. But it must be a continued effort.

For the literate classes there are the newspapers and other publications as media for propaganda; and for all classes there is the cinema screen.

Suitable slogans could be devised and shown as a routine matter at commencement and during intervals of all cinema shows, accompanied twice a week by a short talk in regional languages.

In 1944 (14-1-44) the Natural History Society resolved that a popular Nature Magazine be published by the Society, and in 1947 (5-6-47) it was decided that simple natural history booklets be issued in the several languages. The magazine idea was held up during the war years but measures to give effect to both resolutions are now in progress.

Moral support of the Government is essential and financial aid a necessity.

A BRIEF FOR ACTION

1. A decision by the Governments.
2. Issue of a general law to prohibit sale, possession, marketing of meat, hides, horns, etc., of indigenous animals and of birds.
3. Enforcement of Arms Licence rules and conditions.
4. Enforcement of laws and rules under Act VIII of 1912 and Act XIV of 1927.
5. Formation of a Wild Life Department.
6. Propaganda.
7. Generally all possible steps towards saving wild life.

Through the continued efforts of their leaders the peoples of India were roused to political consciousness. Through their long sustained efforts they attained political freedom. Will the leaders and the people not now demonstrate to other civilized nations that they are equally capable of preserving wild life for posterity? Surely they will, Because they should, and because it is demanded for the prestige of India

It was the intention of the Society and the writer to submit this pamphlet to Mahatma Gandhi with appeal for his powerful advocacy. Alas! it was not so ordained.

Yet, in view of the late Mahatma's well known sympathy with all things created, it may surely be hoped that the peoples of India and of Pakistan will respond to this appeal in accordance with what would without doubt have been his wishes and his guidance for the preservation of wild life in this country.

■ ■ ■

DEEPAK APTE

Time is endless in thy hands, my lord.
There is none to count thy minutes.
Days and nights pass and ages bloom
and fade like flowers.
Thy centuries follow each other
Perfecting a small wild flower

— Tagore, Gitanjali